Bioenergetics is the name given to the collection of disciplines within biochemistry and physiology that aims at a description and understanding of the function of living systems. These functions include the exchange, metabolism and accumulation of matter as well as their related energy transformations.

The authors present an accessible textbook providing the reader with the fundamental principles of the subject and how these can be applied to practical problems.

This textbook is ideal for graduate students and researchers in biochemistry, physiology and chemistry. The book should also be of interest to workers in the applied life sciences.

Bioenergetics

Bioenergetics:

its thermodynamic foundations

LARS GARBY
Department of Physiology, University of Odense, Denmark

and

POUL S. LARSEN
Department of Fluid Mechanics, Technical University of Denmark

CAMBRIDGE
UNIVERSITY PRESS

CAMBRIDGE UNIVERSITY PRESS
Cambridge, New York, Melbourne, Madrid, Cape Town, Singapore, São Paulo

Cambridge University Press
The Edinburgh Building, Cambridge CB2 8RU, UK

Published in the United States of America by Cambridge University Press, New York

www.cambridge.org
Information on this title: www.cambridge.org/9780521451437

English translation © Cambridge University Press 1995

This publication is in copyright. Subject to statutory exception
and to the provisions of relevant collective licensing agreements,
no reproduction of any part may take place without the written
permission of Cambridge University Press.

First published in Danish as *Bioenergetik*
by G.E.C. Gad Publishers, Copenhagen 1991,
and © G.E.C. Gad Forlags 1991

First published in English by
Cambridge University Press 1995 as *Bioenergetics*

This digitally printed version 2008

A catalogue record for this publication is available from the British Library

Library of Congress Cataloguing in Publication data

Garby, Lars.
    Bioenergetics: its thermodynamic foundations / Lars Garby and
Poul S. Larsen.
        p. cm.
    Includes index.
    ISBN 0-521-45143-4
    1. Bioenergetics. 2. Thermodynamics. I. Larsen, Poul Scheel.
II. Title.
QP517.B54G37 1995
574.19′121—dc20 94-956 CIP

ISBN 978-0-521-45143-7 hardback
ISBN 978-0-521-06635-8 paperback

I advanced some time ago the general hypothesis that in consequence of the fluid state of the protoplasm and the instability of the cell-stuffs, voluntary events of physical and chiefly of chemical nature are going on continuously which aim at a balance of the existing potentials of energy. Since life requires the continuation of these potentials of energy, work must be performed continuously for the prevention or reversion of these spontaneous changes.

(Otto Meyerhof, 1924)

Contents

Preface

Bioenergetics is the name given to the collection of disciplines within biochemistry and physiology that aims at a description and understanding of the function of living systems, i.e. collections of whole organisms (ecosystems), individual organisms or parts thereof. The function includes the exchange, metabolism and accumulation of matter (growth) as well as the energy transformations associated with these phenomena. The quantitative description is based on the theory of macroscopic thermodynamics.

Fundamentally, thermodynamic theory is based on the recognition of: the difference between the energy content of a system and energy transfer in the form of heat and work; the difference between the energy forms heat and work; and the existence of an absolute temperature scale and a property of state called entropy. The theory leads to quantitative relations between these phenomena, thereby determining their extent. These results arise as natural consequences of satisfying the first and second laws of thermodynamics (the conservation of energy and the increase of entropy), conditions that are even more far-reaching than, for example, the conservation of mass, which, of course, must also be satisfied.

Existing textbooks and chapters on bioenergetics discuss the fundamental principles only sparsely and the language often does not correspond to that used in textbooks on thermodynamics. This fact has contributed to making the communication between biologists and physicists or engineers unnecessarily difficult. It is also one of the objectives of the present treatise to make this dialogue easier.

The text is intended for students and teachers at schools of higher education with programmes in the fields of biology and medicine, such as biochemistry, physiology, microbiology and ecology, but also for researchers in the applied sciences of, for example animal and human nutrition, food science and plant technology, including fish farming.

Readers who are already familiar with chemical thermodynamics are encouraged to read Chapters 1 and 2 as an introduction to the subject and to the philosophy of the text. They may then skip Chapter 3, which contains general definitions and concepts, an introduction to state properties and the conservation of mass for closed and open systems. The following chapters deal with biological systems in terms of conservation of energy (Chapter 4), increase of entropy (Chapter 5) and expenditure of exergy (Chapter 8). Sections on thermodynamic equilibrium (Chapter 6) and non-equilibrium systems (Chapter 7) are necessary for a complete treatment, particularly for the description of equilibria and transport processes across membranes. Throughout the text, examples illustrate applications of the theory. The property tables in the appendices are not meant to be exhaustive and exact, but they do include sufficient values to be representative and to allow the solution of simple problems.

Equations are numbered consecutively throughout each chapter and are referred to there by their number only, e.g. (72). Reference to equations outside chapters is made by chapter number and equation number, e.g. (3.72).

Our thanks are due to Professor C. Barker Jørgensen and Associate Professor F. Jørgensen for valuable comments.

Lars Garby
Poul Scheel Larsen

1
Introduction

Living systems in the form of whole organisms or parts thereof, down to isolated cells, are extremely complicated. Their function involves a great many processes, and a complete description has not yet been possible. Examples of processes include chemical and electrochemical processes, transport of matter by way of fluid convection or diffusion through solutions and membranes, and transfer of energy to and from the systems.

In order to delimit the concept of bioenergetics and to identify useful analytical tools, we first consider a number of different aspects that can be used for analysing a biological system. Next, we describe the principal functional organization of living systems by way of a simple model, which will be used later in a number of examples. Finally, we mention briefly those tools that are at our disposal for analyses of biological systems.

1.1 Five points of view

Depending on the point of view, a system can be analysed and evaluted with regard to different aspects.

One aspect may be the characteristic function of the system (Figure 1.1). For example, the liver removes a number of toxic substances from the blood and excretes them through separate channels. Certain micro-organisms move with the help of their cilia through water, in which there are concentration gradients. The wilful lifting of an arm is brought about when the necessary nerve impulses reach the relevant muscles. One individual maintains a certain body weight at a given food energy intake, while the same intake leads to an excess of weight in another individual. Such considerations are qualitative and are points of origin for quantitative considerations described in the following.

Turning to quantitative aspects, consider the exchange of matter

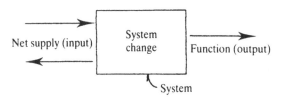

Fig 1.1 The function of the system.

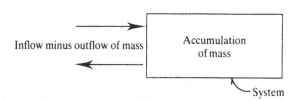

Fig 1.2 The conservation of mass.

between a system and its environment (Figure 1.2). How does the mass of a cow change over time, given an intake (feed, water, etc.) and an output (milk, excretions, etc.) of matter? The relationship between these quantities is given by the law of conservation of mass.

A third aspect is the exchange of energy with the surroundings (Figure 1.3). How much energy has to be expended by a human individual in order to maintain a given body temperature and perform a given amount of external work? What is the energy expenditure for maintenance of the basal functions of a given tissue? The answers to these questions require the use of the law of conservation of energy, i.e. the first law of thermodynamics. This, in turn, requires knowledge of the mass fluxes into and out of the system, the chemical reactions within the system and the thermodynamic properties of the substances involved, such as the specific energy of each component mass.

A fourth aspect addresses the quality of the system and the processes taking place within it (Figure 1.4). A refrigerator, for example, can be characterized by a measure of quality which depends on how well it is insulated from the surroundings and on the efficiency of its heat pump, i.e. the ratio between the heat removed from the cooling compartment and the power used by the pump. The refrigerator can, furthermore, be characterized by its 'thermal structure', expressed in terms of values of absolute temperatures inside and outside the refrigerator. The pump is required to maintain the difference between these temperatures because

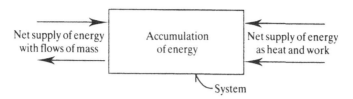

Fig 1.3 The conservation of energy.

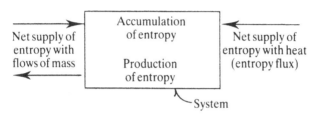

Fig 1.4 The increase of entropy due to irreversibility.

the dissipative process of heat conduction across the walls would otherwise destroy the thermal structure. This example suggests the concept of 'dissipative structures'. Similarly, living systems have structures. Pumping processes are required to maintain these structures in competition with the inherently dissipative processes that would otherwise destroy the structures. Examples of structures are differences in concentrations of Na and K ions across membranes, and states of chemical non-equilibria such as concentrations of proteins that are much larger than those predicted by the actual concentration of amino acids. Such structures require 'chemical pumps' which use energy and which have limited efficiencies. The concept of entropy and the law of increase of entropy in isolated systems, i.e. the second law of thermodynamics, can be used to describe quantitatively the efficiency of maintaining dissipative structures.

A fifth aspect, which combines the third and the fourth aspects, focusses on that part of energy that is called the availability, or the exergy, or the useful energy (Figure 1.5). Systems that are built to deliver high-quality energy in the form of work, or to maintain dissipative structures, depend not only on a net supply of energy but, in particular, on a net supply of exergy. The quality of a system depends on the degree of dissipation, as expressed in the irreversibility within the system. The irreversibility is a measure of the destruction of exergy, and this term

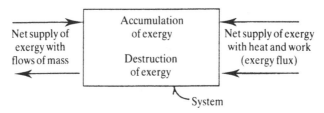

Fig 1.5 The expenditure of exergy due to irreversibility.

appears in the balance of exergy, which is a law that combines the first and second laws of thermodynamics.

The preceding five view points applied to living systems form the basis for bioenergetics. In other words, bioenergetics seeks, firstly, to establish a qualitative understanding of the functional organization of a system. Secondly, it tries to quantify the description by using the tools of physics, chemistry and thermodynamics. The quantitative descriptions determine fluxes, and changes in the mass, energy, entropy and exergy of the system. Such quantitative descriptions must be consistent and they will supplement the description of the functional organization of the system (Garby and Larsen 1984).

1.2   The functional organization of living systems

From a thermodynamic point of view, the functional organization of living systems can be described, in principle, as shown in Figure 1.6.

In the animal kingdom, energy is supplied with the intake of food stuffs. Various chemical compounds (carbohydrate, fat and protein) react with oxygen in reaction (1) to form products (carbon dioxide, water and nitrogen-containing small molecules) that are excreted. In plants, on the other hand, the primary source of energy is the radiant energy from the sun which is used in photosynthetic processes along with absorbed carbon dioxide, water, etc. In both cases, part of the energy is converted into a spatial separation of electrons and protons that results in the flux of protons across a specific membrane. Here, the reaction, adenosinediphosphate (ADP) + phosphate → adenosinetriphosphate (ATP), is forced in the direction to the right. The remaining energy is converted into heat and lost to the surroundings for a system operating at constant temperature. Products of reaction (2′), i.e. triglycerides, glycogen and proteins that exist as energy depots, are also combusted in reaction (1).

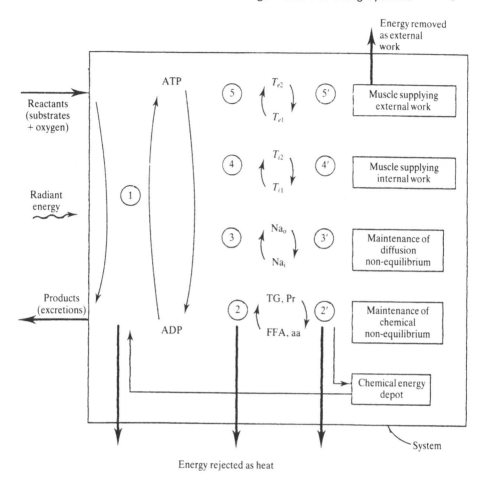

Fig 1.6 The functional organization of living systems. Energy input in the form of radiation is used in photosynthesis in plants. Symbols of reactions include amino acids (aa), free fatty acids (FFA), protein (Pr), triglycerides (TG), intra- and extracellular sodium ions (Na_i, Na_o) and muscle tension for internal and external work (T_i, T_e).

The breakdown of ATP to ADP and phosphate is coupled to a large number of energy consuming processes of which the four main routes are shown. The couplings are associated with the production of heat.

In reaction (2), the $ATP \rightarrow ADP$ process is coupled to processes of formation of protein molecules (Pr) from amino acids (aa), triglycerides (TG) from free fatty acids (FFA) and glycogen from glucose. A continuous breakdown of these molecules in reaction (2') ensures the maintenance of a stationary state of chemical non-equilibria. Triglycerides and glycogen, in particular, can be temporarily stored in depots.

In reaction (3), the ATP→ADP process is coupled to the transport of ions, mainly Na^+, K^+ and Ca^{++}, across membranes. Simultaneous passive back-diffusion of these ions in (3′) maintains non-equilibria of the ions, a phenomenon of fundamental importance for cellular functions.

In reactions (4) and (5), the ATP→ADP process is coupled to the contraction of muscle fibres so that mechanical work can be performed. These processes are symbolized by the establishment of time-averaged tensions, $T_2 > T_1$. Internal mechanical work (4′), involves the pumping of blood by the heart, movement of air by the lungs, chewing and swallowing, and the peristaltic function of the bowel, etc. Similarly, external mechanical work (5′) is performed in movements necessary for walking, swimming and flight, and for manual work. It is clear that external mechanical work cannot take place without performing some internal mechanical work.

With the exception of the external work in (5′) and the possible storage of energy, all the energy uptake by the system is eventually lost as heat. The heat loss from reactions (1)–(5) is an expression of limited efficiency in the couplings, while that from reactions (2′)–(5′) is due to their dissipative nature.

The total system, or its parts, in Figure 1.6 can be considered from each of the points of view shown in Figures 1.2–1.5 and so can be subjected to detailed bioenergetic analysis.

1.3   Objects and limitations

It is our objective to present the tools necessary for a detailed bioenergetic analysis of a given system. These tools include (i) a well-defined terminology for concepts, (ii) a systematic way of treating problems, (iii) the mathematical formulation of the laws of nature and (iv) the properties of matter necessary for computing quantitative results.

Since living systems are extremely complicated and since many of their details are still under investigation, the analysis of real systems is possible by introducing idealized simplifications. For these cases, however, it is possible to write down balances for mass, energy and exergy, and to make statements about the quality of the system and its processes.

However, there remains an important element in the description of systems. As mentioned above, it is a characteristic of any living system that its composition and state, expressed by a very large number of quantities (concentrations, pressure, temperature, etc.), remains essentially constant over time, despite internal and external disturbances. This fact requires the existence of a very large number of control systems, each consisting of (i) sensors or detectors with their input and output signals; (ii) a centre, or set-point, for comparison with reference signals; and (iii) effector organs for the delivery of the necessary effects to keep set-points. These control systems are typically interconnected so that the output, i.e. the regulated quantity, of one system is an input in another system. An example is the combined control system for maintaining the body energy content as well as the body temperature, both of which must be assumed to be regulated quantities and to have common effector elements. The processes involved in the detection and transmission of the necessary signals are, however, in general believed to require very small amounts of energy and they cannot, in general, be traced by bioenergetic analysis.

2

Outline of thermodynamic theory

The formulation of macroscopic thermodynamic theory can be summarized in the general laws of thermodynamics (zeroth, first, second and third laws), supplemented by a number of particular laws (equations of state and the concept of equilibrium). The formulation gives a quantitative description of natural phenomena associated with the equilibrium state of material systems, their energy and energy exchange.

By general laws we understand natural laws which are independent of the particular constitution of the medium in question. Particular laws, on the other hand, depend on the medium. In addition, we generally assume the state to be uniform and at equlibrium in a given medium, which is then homogeneous.

These fundamental considerations are summarized in the present chapter, which also gives a brief summary of the historical development of thermodynamic theory. Finally, some practical advice is given with respect to the formulation and solution of problems.

2.1 Presuppositions

The formulation of the theory is based on a number of assumptions that, to some extent, are implicit in the statement of natural laws:

- A material system is assigned a mass M and a total energy E. The energy is the sum of the internal energy U, kinetic energy KE and potential energy PE:

$$E = U + KE + PE. \tag{1}$$

- Energy can be added to a system in the form of work δW and in the form of heat δQ, where δ refers to a small (differential) quantity.

- The mechanical (kinetic or potential) energy of a system can be changed by transferring energy in the form of work due to the displacement of an external force. The electrical potential energy of the system can be changed in a similar way.
- If changes in kinetic and potential energy are excluded and if the system receives energy in the form of reversible work δW_{rev}, this will occur in a process associated with changes in some characteristic extensive parameters of the system. The differential work can then be expressed as

$$\delta W_{rev} = \Sigma F_j \, dX_j, \tag{2}$$

where F_j denotes the generalized intensive parameters (forces) associated with the generalized extensive parameters (displacements) X_j, $j = 1, 2 \ldots$.
- If the system receives energy in the form of reversible heat δQ_{rev}, its energy is increased without changes in the extensive parameters X_j.
- A homogeneous system can exist in an equilibrium state, which can be uniquely described macroscopically by the value of its internal energy and the values of its parameters X_j.
- In an equilibrium state, the entropy S of the system exists as a unique function of the parameters of the state:

$$S = S(U, X_j). \tag{3}$$

This is the state postulate. It implies that the number of independent state properties, which uniquely determine the state, is equal to one plus the number of reversible forms of work.

2.2 General laws

These laws can be perceived as observed natural laws and they can be expressed as follows:

Zeroth law. *Equality of temperature.* Two systems A and B, each of which is in thermal equilibrium with a third system C, will also be in mutual thermal equilibrium and have the same temperature:

$$T_A = T_C \text{ and } T_B = T_C \text{ implies } T_A = T_B. \tag{4}$$

First law. *Conservation of energy.* For a control mass, which by definition is a material system with constant mass, the increase in total energy dE is the sum of supplied energy in the form of heat δQ and work δW:

$$dE = \delta Q + \delta W. \tag{5}$$

Second law. *Increase of entropy.* The increase of entropy dS in a control mass is never less than the added energy as heat δQ divided by the absolute temperature of the mass:

$$dS \geqslant \delta Q/T. \tag{6}$$

For a reversible process, the increase dS is equal to dQ_{rev}/T. For a non-homogeneous system, (6) is replaced by $dS \geqslant \delta(Q/T)$, as known from continuum mechanics, but also recognized in general (Arpaci, 1986).

Third law. *The zero point of entropy (Nernst postulate).* The entropy of pure matter is zero at the zero point of the absolute temperature scale:

$$\lim_{T \to 0} S = 0. \tag{7}$$

Whenever the thermodynamic system includes mechanical, electrical and/or magnetic phenomena, general laws for each of these branches of physics must be added. For example, when thermomechanical systems are being analysed, we use the following from mechanics:

Continuity. *Conservation of mass.* The increase in total mass M of a control mass is zero:

$$dM = 0. \tag{8}$$

Newton's second law. *Balance of momentum.* The product of the mass of a control mass M and its linear acceleration dV/dt is equal to the sum of forces K which act in the direction of the acceleration:

$$M \, dV/dt = K. \tag{9}$$

The general laws and underlying assumptions described above can be used to derive the Gibbs relation, the maximum of entropy of an isolated system, and other thermodynamic relations and conditions for equilibrium and stability. Furthermore, they can be used to prove the impossibility of perpetual motion of the first and the second kind.

The statements made up to now are valid for a simple homogeneous system consisting of one component only, as well as for a homogeneous system that consists of a non-reactive mixture of several components (chemical species). Of course, the latter case needs the state properties of mixtures. On the other hand, analysis of chemically reactive mixtures requires the formulation to be augmented by equations which prescribe the possible chemical reactions and equilibria. These equations constitute, like the general assumption of equilibrium, compatibility conditions to which the solution of the given problem is subjected.

It should be noted that the present treatment of the foundations of thermodynamic theory has been selected to present, in a straightforward manner, analytical tools that are available for the solution of practical problems. It may be shown that there is redundancy in the exposition. However, our task is not to expose the theory in the shortest and most concise way in one or two abstract postulates in order to reach the same results by deduction. Interested readers are referred to other texts, e.g. Carathéodory (1909), Brønsted (1955), and Hatsopoulus and Keenan (1962, 1965).

2.3   Historical outline

In physics, we are confronted with a number of phenomena which form the basis for the formulation of general and particular natural laws. From special cases, in the form of real or imaginary experiments, induction leads from a part to the whole and thus to general laws that govern the physical phenomena that we observe. These laws have been summarized in Sections 2.1 and 2.2 and they form the starting point for the following treatment.

Seen in retrospect, the formulation of thermodynamic theory has been associated with difficulties. We have had to recognize the difference between the content of the energy of a system, on the one hand, and the energy forms work and heat, on the other, the former being energy associated with matter, the two latter being energy in transit; furthermore we have had to recognize and describe qualitatively and quantitatively the difference between heat and work as transferred forms of energy, an absolute temperature scale and, finally, a state property called entropy. The four concepts: work, heat, absolute temperature and entropy, are, in their relation to the concept of energy, fundamental for macroscopic thermodynamic theory. As a matter of interest, we note that during the development of thermodynamic theory – just as in electromagnetic theory – there was considerable confusion because no clear distinction was made between fundamental principles (general laws) and constitutive relations (particular laws). The list of nineteenth-century scientists who contributed to thermodynamic theory is long and includes such names as Fourier, Carnot, Clapeyron, Helmholtz, Kelvin, Rankine and Clausius. However, it was Gibbs in particular, who, as a mathematician, contributed to the clear formulation and exposition of relations between thermodynamic functions. Various views on the historical development may be found in Carathéodory (1909), Gibbs (1948), Brønsted (1955), Pledge (1959), Hatsopoulos and Keenan (1962, 1965), Giedt (1971) and Truesdell (1971).

The recognition and formulation of the first and second laws was closely related to the invention of the steam engine and the combustion engine and the study of the special cyclic processes of these machines. It was realized that the machines could produce energy in the form of mechanical work from the heat supplied by combustion and transferred to the working medium of the machines. Conservation of energy, expressed in the first law, was immediately understood as soon as work and heat were recognized to be energy in transit and equivalent in some sense when measured in the same energy units. The fact that all the energy supplied as heat could not be transformed and delivered as work was expressed in the second law. It was appreciated that a working medium, say in the form of a gas at a high pressure, even in an ideal machine, could only be expanded to the pressure of the surroundings (and not to zero pressure). Also, the temperature of the medium undergoing a spontaneous process could only be decreased to the temperature of the surroundings (and not to the absolute zero of the temperature scale). Concepts such as maximal work from a cyclic machine, useful energy (exergy) for a given state of a medium, and possible and spontaneous processes, could now be easily explained in terms of the second law and the concept of entropy.

Because of its historical origin, classical thermodynamics came to be expounded from the rather narrow view of treating state changes in simple compressible media, i.e. media with compression as the only reversible form of work. The interest in combustion processes and other chemical reactions led to the development of chemical thermodynamics and the analysis of mixtures, phase equilibria, reactive mixtures and chemical equilibria, a development motivated mainly by the chemical process industry. The development did not end there and we now recognize the generality of thermodynamics, including thermomechanical, electrical, magnetic and chemical energy concepts and associated changes of state. It soon became evident that biological systems, although much more complicated than man-made systems, obey the same natural laws, and the application of thermodynamics was also extended to this field. Thus, with advances in biochemistry and physiology, bioenergetics is increasingly used to study energy requirements and energy expenditures in living systems. This affirms the universality of macroscopic thermodynamics, being a general analytical tool in many disciplines.

The present treatise is based on classical (macroscopic) thermodynamics, even though explanations and arguments sometimes refer to the microstructure of media and thus use elements of statistical (microscopic) thermodynamics. Classical thermodynamics deals, strictly speaking, only with homogeneous systems at global equilibrium. However, a non-homogeneous system can be divided into subsystems, each of which may be regarded as a homogeneous system at local equilibrium. In this case, transport processes (of heat, matter and electrical charge) can be regarded as discontinuities with specified fluxes and potential differences between the subsystems. The continuum formulation appears in the limits of differential subsystems. Here, state properties, fluxes and potentials vary continuously throughout the system. This approach is used in the analysis of systems at global non-equilibrium, usually with the assumption of local equilibrium. The approach is generalized with reference to irreversible thermodynamics, which describes fluxes phenomenologically in terms of differences or gradients in the driving potentials.

2.4 Formulation and solution of problems

Mastering a theoretical discipline, such as bioenergetics, can be summarized in three elements: knowledge of the terminology with its underlying concepts, definitions, etc.; the cumulation of experience from examples and problems, relating terminology and concepts to models of actual processes; and, finally, the conscious understanding of – and ability to use – the analytical tools of the discipline. Mastering the latter is facilitated by logical structuring of the knowledge, as suggested in the introduction to this chapter.

We suggest the following systematic approach to the *formulation* of a specific problem:

1. Make a schematic *drawing* of the components of the problem, showing all relevant details: mass flows, energy flows, energy fluxes as heat and

work, composition of mixtures, chemical reactions, given states in the form of temperature, pressure, concentration, electrical potential, etc. Try to illustrate which and how processes proceed. Are they steady or unsteady, isothermal, isobaric, etc?

2. Define (i.e. choose) the *system* to be analysed by drawing a control surface that delimits the system (control mass or control volume). Which of several possible material systems should be the object for analysis? Which are the flows and fluxes that cross the control surface? Are they known or should they be determined? In fact the chosen control surface determines the quantities that will appear in the formulation of the resulting equations. In many problems it may be necessary to analyse several systems separately, e.g. a part of a living system and its surroundings.

3. Write the relevant *general laws* of the system. Use the drawing of the system to identify the contributions from the volume and from the surface to these laws.

4. Write the relevant particular laws, i.e. *compatibility conditions*, such as the Gibbs relation, balanced reaction equations and equilibrium conditions, and *constitutive relations*, in the form of equations of state for the calculation of state properties or in the form of values read from tables or charts.

5. Note the given, obvious or simplifying auxiliary conditions or assumptions which make the formulation *complete* (the number of unknowns for a unique solution must equal the number of equations).

Figure 2.1 summarizes the elements of the mathematical formulation that may be needed for a general system involving reactive mixtures.

Following the mathematical formulation of a problem comes its *solution*. Although this phase should not present any difficulty, the following comments may be useful:

1. Reduce the mathematical formulation without introducing numbers for symbols and make justified mathematical simplifications or approximations.

2. Check, by way of a dimensional analysis, that all terms in an equation have the same dimensions.

3. Insert values using a consistent (homogeneous) system of units. Judge whether the result appears to be reasonable. Make a rough order-of-magnitude estimate.

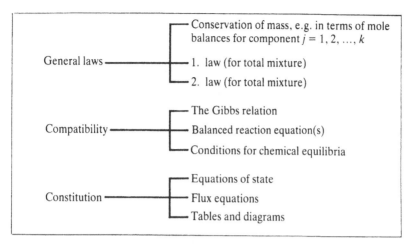

Fig 2.1 Elements of the formulation.

The approach suggested here will be demonstrated explicitly in some of the examples that follow.

3

Concepts, definitions and state properties

In this chapter we summarize a number of fundamental concepts and definitions which, in regard to understanding and terminology, set the framework for the subsequent treatment. The general law of mass conservation is fundamental and is treated here, first for a simple one-component substance and then for mixtures of several components. The concept of energy, including forms of work and heat transfer, are discussed and illustrated by examples. Other fundamental topics include: state properties, equations of state, and general relations between state properties to the extent that these are used later for pure substances and for mixtures. More complete and systematic treatments can be found in van Wylen and Sonntag (1965) and Rock and Gerholdt (1974). The chapter concludes with some general considerations of processes in living systems.

3.1 Control mass (CM) and control volume (CV)

The analytical formulation of a thermodynamic problem consists *inter alia* of application of the general laws (notably the conservation of mass and the first and second laws) to a given quantity of matter. In order to define this quantity of matter precisely, we introduce the concept of a *control surface*. This is a real or virtual closed surface which delimits the quantity of matter being considered. The control surface can be rigid or deformable, fixed or moving in space, impermeable or permeable to matter and energy in the form of work and heat.

If the control surface is chosen to be impermeable to matter it will contain a constant mass which is called a *control mass* (CM, closed system), see Figure 3.1(*a*).

If, on the other hand, the control surface is chosen so that one or several mass flows cross it, the enclosed space is called a *control volume*

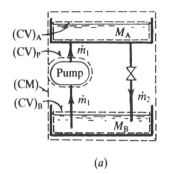
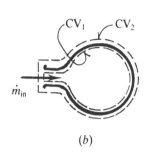

(a)   *(b)*

Fig 3.1 Examples of choices of control mass and control volumes. (a) The pump delivers the mass flow $\dot{m}_1$ to the reservoir with mass M_A, from where mass flow $\dot{m}_2$ returns by gravity to reservoir M_B. Any of the control volumes CV, or the whole system CM, can be chosen for analysis. (b) Deformable control vomume. $(CV)_1$: gas in balloon. $(CV)_2$: gas plus balloon wall. The pressure on the $(CV)_2$ surface equals the pressure of the surrounding atmosphere.

(CV, open system), see Figure 3.1(*a*,*b*). Everything external to the control surface is denoted as the *surroundings*. The term *system* is often used for both control mass and control volume when the difference between them has no significance.

Biological systems are open since they exchange matter with their surroundings, and should be treated as control volumes. Transport of matter within these systems takes place by flow (food intake, movement in the digestive, respiratory and circulatory systems, etc.) when distances are large and by diffusion (transport between and within cells) when they are small.

In general a control surface can include one or several components. Thus, (CM) in Figure 3.1 is a *composite system* consisting of a pump and two reservoirs, each of which can be considered as a control volume. Such a subdivision of a composite system into several control volumes may facilitate thermodynamic analysis and provide additional information.

The system under consideration can often be considered to be in a *spatially uniform state*, or, simply, a *uniform state*. For example, the density ϱ_A of the liquid in reservoir A in Figure 3.1(*a*) can be assumed to be the same throughout so that $M_A = \varrho_A \mathcal{V}_A$, where $\mathcal{V}_A$ is the volume of the liquid. If the state is not spatially uniform, the system can be divided into subsystems, each of which can be assumed to be in a uniform state.

Table 3.1. *Some extensive properties of matter*

	B (extensive)	$b = B/M$ (specific extensive)
Volume	$\mathscr{V}$	v
Mass	M	1
Momentum (linear)	MV	V
Energy	$E = Me$	$e = u + \frac{1}{2}V^2 + \varphi$
Entropy	$S = Ms$	s

$E = U + KE + PE$ is the total energy; $u = U/M$ is the internal energy, $\frac{1}{2}V^2 = KE/M$ is the kinetic energy, $\varphi = PE/M$ is the potential energy, and V is the velocity (momentum per unit mass).

The sum (or, in the limit of continuously distributed systems, the integral) of contributions of the subsystems describes the total system.

Processes in a system can be steady or unsteady. By definition, a *steady* process satisfies

$$dB/dt = 0, \tag{1}$$

where B denotes any of the extensive properties describing the state of the system (Table 3.1). Thus, if M_A and $\mathscr{V}_A$ are constant for $(CV)_A$ in Figure 3.1, we speak of a steady process, even though liquid is flowing in and out of the reservoir.

It is sometimes useful to choose a deformable control volume, such as the volume of gas contained at any time in a balloon under expansion (Figure 3.1(b)). As indicated in the diagram, the balloon wall can be excluded $(CV)_1$ or included $(CV)_2$; this will be reflected in the formulation of the problem since the mass, energy, etc., of the control volume will be different in the two situations. In the case where $\dot{m}_{in} \neq 0$, the process is unsteady.

3.2 Balance equations for control mass and control volume

A general law can be expressed as a balance equation. This equation for a control mass can be transformed into an equation for a control volume by the use of the Reynolds transport theorem, which establishes the relation between the change over time of a property in the two systems.

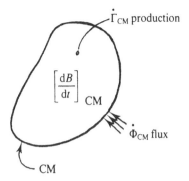

Fig 3.2 Balance equation for a control mass.

3.2.1 Balance equation for a control mass

Each of the general laws, which are natural laws originating from observations, can be written in the fundamental form of a *balance equation for a control mass*:

$$[dB/dt]_{CM} = \dot{\Phi}_{CM} + \dot{\Gamma}_{CM}. \tag{2}$$

This equation states that the change over time of the property B within the control mass equals the sum of the flux $\dot{\Phi}_{CM}$ of B through the control surface and the production $\dot{\Gamma}_{CM}$ of B within the volume (Figure 3.2). Here, and later in the text, the dot ($\cdot$) over a symbol denotes the quantity of the property in question per unit of time that is associated with a transport (e.g. flow or flux through the control surface) or with a production (inside the control surface). The notation does not refer to a derivative with respect to time in the usual sense.

Many problems of practical importance involve mass flows through a process component and it may be difficult or impossible to analyse such problems by the use of (2), which is valid for a fixed mass. It would be much simpler to consider a control volume that surrounds the process component, allowing inflow and outflow of mass. We therefore need the form of (2) for a control volume.

3.2.2 The Reynolds transport theorem

Figure 3.3 shows a control mass and a control volume in a flowing medium. The control surfaces of the two systems coincide at time t_0 when the two systems enclose the same quantity B_0 of the property B. At a

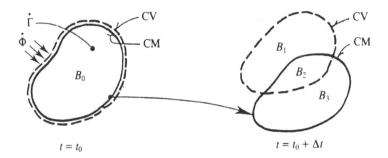

Fig 3.3 Relation between control volume and control mass.

later time $t_0 + \Delta t$, the control surfaces contain different masses because the control mass must follow the movement of the medium, while the control volume can be chosen to be fixed in space or in arbitrary motion with or without a change in its volume. While the quantity B_2 is common to the two systems the quantity B_3 represents an outflow and the quantity B_1 an inflow through the surface of the control volume during time Δt. This means that the time-rate-of-change of the property B for $\Delta t \to 0$ is different for the control mass and the control volume, being related by

$$[dB/dt]_{CM} = [dB/dt]_{CV} + (\text{outflow minus inflow})_{CV}. \qquad (3)$$

The last term in (3) can contain several contributions of the form

$$\text{outflow of } B = (\dot{m}b)_{out}, \text{ inflow of } B = (\dot{m}b)_{in}, \qquad (4)$$

where $b = B/M$ is the specific property of the substance and $\dot{m}$ denotes the instantaneous value of the mass flow crossing the surface of the control volume at time $t = t_0$. Substitution of (4) into (3) gives *the Reynolds transport theorem*

$$[dB/dt]_{CM} = [dB/dt]_{CV} + \Sigma(\dot{m}b)_{out} - \Sigma(\dot{m}b)_{in}. \qquad (5)$$

This equation states that the change over time of the property B within the closed system (control mass CM) can be expressed in terms of quantities pertaining to the open system (control volume CV) as the change over time of B plus contributions from the net outflow of B.

This result presupposes a uniform distribution of the states in the outflows and the inflows, of which there can be many, as indicated by the summation signs. When the states are distributed non-uniformly, the total amount of B in the given system is obtained by summation or integration.

3.2.3   Balance equation for a control volume

Using (5) on the left-hand side of (2), we obtain the general form of the *balance equation for a control volume*:

$$[\mathrm{d}B/\mathrm{d}t]_{\mathrm{CV}} + \Sigma\,(\dot{m}b)_{\mathrm{out}} - \Sigma\,(\dot{m}b)_{\mathrm{in}} = \dot{\Phi}_{\mathrm{CV}} + \dot{\Gamma}_{\mathrm{CV}}. \tag{6}$$

Since, at time $t = t_0$, the control volume and control mass are identical, the flux $\dot{\Phi}$ of B through the control surface and the production $\dot{\Gamma}$ of B within the volume are the same for the two systems. This, however, assumes that the convective mass flow through the surface does not contribute to the flux. Later, when the first law for a control volume is derived (see Section 4.2), it is shown that flow in conjunction with pressure forces will give an additional contribution to the flux in the form of mechanical work per unit time.

In the following, we will always write general laws in the form of (6) and make a clear distinction between flux and flow. *Flux* is defined as transport of matter by way of diffusion, and *flow* is defined as the transport of matter by way of motion (convection).

3.3   Conservation of mass

Conservation of mass (2.8) for a general unsteady process in a control mass becomes, after dividing by the differential time interval $\mathrm{d}t$ and the usual limit process,

$$\mathrm{d}M/\mathrm{d}t = 0 \ (\mathrm{CM}). \tag{7}$$

This is (2) with $B = M$ and $\dot{\Phi} = \dot{\Gamma} = 0$. The definition of control mass excludes the flux of mass through the control surface, and conservation of mass excludes production in the volume.

For a process $1 \rightarrow 2$, integration of (7) over the elapsed time gives

$$M_2 - M_1 = 0 \ (\mathrm{CM}). \tag{8}$$

showing, as expected, that the mass at state 2 equals that at state 1.

Employing the Reynolds transport theorem (5), with $b = 1$, in (7) immediately gives the conservation of mass for a control volume:

$$\mathrm{d}M/\mathrm{d}t + \Sigma\,\dot{m}_{\mathrm{out}} + \Sigma\,\dot{m}_{\mathrm{in}} = 0 \ (\mathrm{CV}). \tag{9}$$

The same result is obtained from (6) with $B = M$, $b = 1$ and a right-hand side, which, as before, is equal to zero. The result (9) is the mass balance of open systems, see Figure 1.2. The form is valid for a pure substance or a mixture. For a mixture, mass and mass flows refer to the total mixture. The mass balance can be expressed in terms of mole numbers, as shown in Section 3.9.2, which also gives the form for a single component of a mixture.

For a steady process, $dM/dt = 0$, and (9) shows that the sum of mass flows out equals the sum of mass flows in. If no mass flows through the surface of the control volume, (9) reduces to (7), as expected, and the control volume becomes in fact a control mass. Hence, form (7) is a special case of the general form ((9)).

Example 3.1

For the four systems shown in Figure 3.1(a), the conservation of mass (9) can be written as follows, when the masses of liquid in the pipe and pump are constant or negligible:

$$d(M_A + M_B)/dt = 0 \text{ (CM)}, \tag{a}$$
$$dM_A/dt + \dot{m}_2 - \dot{m}_1 = 0 \text{ (CM)}_A, \tag{b}$$
$$\dot{m}_1 - \dot{m}_1 = 0 \text{ (CV)}_P, \tag{c}$$
$$dM_B/dt + \dot{m}_1 - \dot{m}_2 = 0 \text{ (CV)}_B. \tag{d}$$

As expected, the sum $(b) + (c) + (d)$ gives (a).

3.4 Energy

A material system contains mass M and total energy E, being the sum of internal, kinetic and potential energies, according to (2.1):

$$E = U + KE + PE = U + \tfrac{1}{2} MV^2 + M\varphi, \tag{10}$$

where φ is the potential energy per unit mass. The material system can be a control mass, a control volume or a part of a mass flow into or out of a control volume.

Energy that is transferred to or away from a system without being associated with matter is identified either as heat or as work. According to this definition, a system does not contain heat energy or work energy, but only energy associated with mass according to (10). However, depending on the state of the system in relation to its surroundings, it can receive or give up energy in the form of heat or work through interaction with the surroundings. In other words, heat (including radiation) and work are *energy in transit*. Once such energy has been added to a system, its origin cannot be traced.

3.4.1 Kinetic energy

The kinetic energy of mass M in linear motion with velocity V relative to a reference point is

$$KE = \tfrac{1}{2}MV^2. \tag{11}$$

The kinetic energy is a state property. Its specific value (i.e. per unit mass) is $\tfrac{1}{2} V^2$. Its magnitude is arbitrary since the reference point can be chosen arbitrarily. This fact is of no importance, however, since only changes in kinetic energy are of interest.

The kinetic energy of a rigid body of finite extent with mass M, moving linearly with velocity V, is also given by (11). For more complicated situations, the reader is referred to texts on mechanics (see e.g. Benedek and Villars, 1974).

Example 3.2

A lion weighing 150 kg and moving with a velocity of 50 km/hour (about 14 m/s) reduces its speed to 0 m/s on collision with a much heavier prey. The decrease in kinetic energy in the inelastic collision is

$$\tfrac{1}{2}M(V_1^2 - V_2^2) = \tfrac{1}{2} \times 150 \times 14^2 = 14\ 700\ \text{J} = 14.7\ \text{kJ}. \tag{a}$$

By comparison, the heat production of the lion at rest is about 150 W or 15 kJ during a period of 100 seconds (Figure 3.4).

V_1 M $V_2 = 0$

Before   After

Fig 3.4

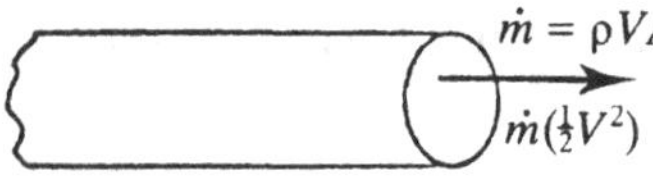

$$\dot{m} = \rho V A \qquad \text{(flow of mass)}$$

$$\dot{m}(\tfrac{1}{2}V^2) \qquad \text{(flow of kinetic energy)}$$

Fig 3.5

Example 3.3

In the human aorta, which can be considered as a tube with an inner diameter $d = 25\,mm$, the blood flow is about $5\,l/min$. The mean velocity is

$$V = \dot{V}/A = 4/\pi d^2 = 4 \times (5 \times 10^{-3}/60)/(\pi \times 0.025^2) = 0.17\,m/s, \quad (a)$$

and the mass flow, using the density of water $\rho = 1000\,kg/m^3$, is

$$\dot{m} = \rho V A = \rho \dot{V} = 0.083\,kg/s. \tag{b}$$

The flow of kinetic energy in the blood stream (Figure 3.5) is

$$\dot{m}(\tfrac{1}{2}V^2) = 0.083 \times 0.5 \times 0.17^2 = 0.0012\,W = 1.2\,mW. \tag{c}$$

This amount is negligible compared to the flow of energy that is associated with the increase in pressure delivered by the heart as a pump (see Example 3.18).

3.4.2   Potential energy

Consider the work performed by a conservative force field (potential field) on a system of mass M. The increase in potential energy of the system is defined as the negative of this work. The gradient (spatial change) of the potential energy is therefore equal to the negative of the force, which is expressed, per unit mass, as

$$d\varphi/dz = -K/M. \tag{12}$$

The field of gravity of the earth is a potential field. Consider a mass M being lifted from z_0 to $z_0 + dz$ in this field having acceleration due to gravity g (force per unit of mass), acting in the negative direction of the z-axis (Figure 3.6(a)). The force times displacement, being the work done by the external force $K = -Mg$, equals the increase in potential energy of the mass:

$$\delta W = K\,dl = -Mg\,dz = M\,d\varphi. \tag{13}$$

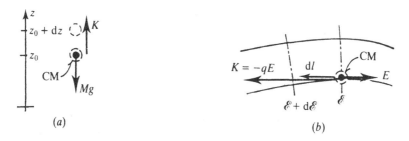

Fig 3.6 (a) Field of gravity; mechanical potential energy. (b) Electric field; electrical potential energy.

This work is also the negative of the work performed by the potential field. Comparison of (12) and (13) shows that the potential energy per unit mass is

$$\varphi = gz. \tag{14}$$

The magnitude of φ is arbitrary, since the reference value for z can be chosen arbitrarily, but this is of no consequence since only changes are of interest.

Example 3.4

A subject weighing 70 kg climbs a hill to a point 1000 m above the point of departure. The increase in potential energy of the subject is

$$Mg(z_2 - z_1) = 70 \times 9.81 \times 1000 = 700\ 000\ \mathrm{Nm} = 0.7\ \mathrm{MJ}, \tag{a}$$

a quantity that can be compared to the energy expenditure during the climb, which is 2–10 MJ, depending on the terrain.

In analogy with the general law of gravitation, electrical charges interact through electrical forces as expressed by Coulomb's law. At any point in an electrical field, the force on a unit charge q_1 ($= +1$ coulomb) defines the electrical field E. The potential energy per unit of charge, which is the electrical potential $\mathscr{E}$, is related to the force field, according to (12), as

$$d\mathscr{E}/dx = -E, \tag{15}$$

where E is the component of force in the direction of the x-axis. If the charge q is moved over a distance dl in an electrical field E, the control mass q receives energy in the form of work (Figure 3.6(b)) of

$$\delta W = K \, dl = -qE \, d\mathscr{E}, \tag{16}$$

since the external force acting on the charge and performing the work is $K = -qE$. This work, which equals the increase in potential energy $q \, d\mathscr{E}$, is the negative of the work performed by the potential field.

An electrical conductor carries a current $\dot{n}_e$ of electrons per unit time, each with a negative elementary charge $-q$ ($= -1.602 \cdot 10^{-19}$ C). The flow of potential energy at a point with electrical potential $\mathscr{E}_1$ is

$$\dot{m}\varphi_1 = \dot{n}_e \, q_e \, \mathscr{E}_1 = -I \mathscr{E}_1, \tag{17}$$

where $\dot{n}_e q_e$ is the negative of the usual electrical current I (C/s = ampère). The magnitude of the electrical potential energy is arbitrary, and only changes are of practical importance.

The foregoing description implies that a system possesses a mechanical potential energy associated with its mass in a gravitational field, and an electrical potential energy associated with its charge in an electrical field. A control volume can receive potential energy with an inflow of mass and an inflow of electrical charge. The potential energy is a state property.

Potentials and potential differences in electrical appliances and circuits are normally induced by external means (e.g. from a dry battery that maintains a voltage difference across a resistor through which flows a current of charge). A solution of electrolytes contains components which are neutral, or positively or negatively charged. Experience shows, however, that a bounded homogeneous solution is electrically neutral since there are no electrical fields within the solution. This is the *principle of electroneutrality*, according to which the number of charges associated with positive and negative ions is practically the same within a given volume.

Electrical potentials and potential differences, not induced by external means, can exist in electrochemical processes across semi-permeable membranes in the laboratory or in living systems. Such potentials are due to processes in the system causing accumulation of charge (see Section 6.4.4).

Changes in electrical potential energy are therefore of particular importance in biological systems that are characterized by charge-carrying ions in solutions whose electrical potentials change, for example across cell membranes. Here, it is convenient to use the potential energy per mole $\hat{\varphi}_i$ of the component at the potential $\mathscr{E}$:

$$\hat{\varphi}_i = z_i \mathscr{F} \mathscr{E}, \tag{18}$$

where z_i denotes the *valency* of the ion (i.e. the number of positive elementary charges q_1 per molecule) and $\mathscr{F}$ is the *Faraday constant*,

$$\begin{aligned}\mathscr{F} = q_1 N_A &= 1.6022 \times 10^{-19} \text{ [C/charge]} \\ &\quad \times 6.022 \times 10^{23} \text{ [molecules/mol]} \\ &\quad \times 1[\text{charge/molecule}] \\ &= 96\,485 \text{ C/mol} \approx 96\,500 \text{ C/mol}, \tag{19}\end{aligned}$$

where N_A is Avogadro's number.

Example 3.5

In general, ions in living systems exist in two phases, the intracellular phase and the extracellular phase. Between these two phases there is a potential difference of about 80 mV, with the intracellular phase being negative. We seek the difference in electrical potential energy per mole positive monovalent ion, e.g. Na^+, between the two phases. From (18), with $z = +1$, we obtain with $\mathscr{F}$ from (19)

$$\begin{aligned}\hat{\varphi}_E - \hat{\varphi}_I = z\mathscr{F}(\mathscr{E}_E - \mathscr{E}_I) &= 1 \times 96\,500 \times 80 \times 10^{-3} \text{ J/mol} \\ &= 7.72 \text{ kJ/mol}. \tag{a}\end{aligned}$$

This quantity is of importance for calculations of the equilibrium distribution of ions (Section 6.4.4) and of the non-equilibrium in relation to ion transport (Sections 4.6 and 7.2.2).

3.4.3 Internal energy

The part of the total energy of a material system that is not associated with the ordered kinetic energy of its mass or with the potential energy of its mass and charge is called the internal energy U of the system.

The internal energy is a state property. For many practical applications of chemical non-reactive systems, its magnitude can be considered as arbitrary. On the basis of the third law of thermodynamics (2.7), however, it can be assumed that the energy of a pure substance is zero at zero absolute temperature. This statement agrees with a commonly used statement that the internal energy of a system is a measure of its ability to undergo a process in which energy leaves the system in the form of heat or work. However, the statement addresses the concept of energy conservation and not the concept of internal energy.

The concept of internal energy is phenomenological in macroscopic theory, and we must refer to microscopic theory in order to make it clear how a material system can accumulate added energy (in the form of heat or work) as internal energy or, conversely, how it can reduce its content of internal energy.

We then consider the system as consisting of a very large number of particles. The internal energy U of this system is the sum of all forms of energy associated with the elementary particles (molecules, atoms, nucleons, electrons, etc.) that constitute the system. These forms of energy can, in principle, be thought of as the microscopic kinetic energy associated with the linear, rotational or oscillating motions of the particles or as the potential energy associated with the positions of the particles in the fields of forces between them. In addition, there are bond energies for chemical compounds.

When the internal energy of the system decreases, the microscopic energies mentioned above decrease in ways that are expressed in the laws of statistical (and quantum) mechanics. For example, in a monoatomic gas at normal temperatures, the decrease in internal energy is due to a decrease in the kinetic energy of atoms only. Macroscopically, a decrease in temperature is observed. The kinetic theory of gases implies that the mean velocity of the particles is proportional to the square root of the absolute temperature. In a polyatomic gas, the internal energy can also be reduced by way of a decrease in the internal molecular forms of energy associated with rotation and vibration. A further decrease can take place with the formation of chemical bonds.

Similar considerations can be applied for a liquid or a solid. The important point is that a material medium has a content of internal energy U, which can increase or decrease as energy is added or removed.

3.5 Work

Work appears as energy transferred to a system when an external force K acting on the system undergoes a displacement. The differential work of the displacement dl in the direction of the force is

$$\delta W = K \, dl. \tag{20}$$

As will be shown in the following sections, the concepts of force and displacement can be generalized to include processes other than those of mechanics. The total work of a finite displacement is obtained by integration of (20). By convention, work W is *positive* for energy *added* to

Table 3.2. *Some reversible forms of work* $\delta W_{rev} = F_j dX_j$ *added to a system with state properties* F_j, X_j

Mechanical	
Compression of gas	$-p\, d\mathscr{V}$
Stretching of elastic wire	$\sigma\, d(\mathscr{V}\varepsilon)$
Stretching of liquid surface	$\sigma_s\, dA$
·Electrical	
Polarization	$E\, d(\mathscr{V}P)$
Magnetic	
Magnetization	$\mu_0 H\, d(\mathscr{V}M)$

p is pressure, $\mathscr{V}$ volume, σ tension (force per unit area), ε strain (relative deformation), σ_s surface tension (force per unit length of section in surface), A area, E electric field strength, P polarization per unit volume, μ_0 permeability, H magnetic field strength, and M magnetization per unit volume.

the system. When the external force undergoes a displacement in its direction, the system receives energy. In the opposite case, energy is transferred from the system to its surroundings.

The energy transferred as work is not a state property but depends on the nature of the process. Work can be performed in two different ways, reversibly or irreversibly. A real process approximates a reversible process when the work is performed by infinitesimal displacements with vanishing imbalance between external force and system reaction and with vanishing friction.

3.5.1 Reversible forms of work

In general, energy in the form of reversible work δW_{rev} can be added to a system through a change in one or several of the extensive parameters X_j, $j = 1, 2, \ldots$ that characterize the system:

$$\delta W_{rev} = \Sigma F_j\, dX_j, \tag{2.2}$$

where F_j are the characteristic intensive parameters (forces) that correspond to the extensive parameters (displacements) X_j. The parameters F_j, X_j are state properties and their mutual relation is determined by the equation of state and the process in question. Table 3.2 summarizes a number of reversible forms of work.

3.5.2 Mechanical work

When a tight-fitting piston is displaced in an open cylinder, energy in the form of work is added from the surroundings to the system (cylinder and piston). The movement of the piston is opposed by forces of friction, and the energy added as (frictional) work is irreversible since it cannot be recovered by reversing the process.

When a frictionless piston is displaced in a closed cylinder containing a gas at pressure p, the system (consisting of the gas) receives energy in the form of work which, for a differential displacement is

$$\delta W_{\mathrm{rev}} = -p \, \mathrm{d}\mathscr{V}, \tag{21}$$

where $\mathscr{V}$ is the volume of the gas. This expression is derived from (20) by noting that the external force in quasi-equilibrium with the pressure forces is pA, where A is the area of the piston, and that the displacement $\mathrm{d}l$ in the direction of the force is $\mathrm{d}l = -\mathrm{d}(\mathscr{V}/A)$. In this case the movement involves a quasi-equilibrium process without friction, and the work (of compression) is reversible since it can be regained by reversing the process.

Example 3.6

The difference between a reversible and an irreversible process is illustrated in Figure 3.7.

A membrane that confines the gas in one half of a rigid container is suddenly removed (Figure 3.7(a)). The gas expands until it fills the whole container. The process $1 \to 2$ is irreversible and no energy is transferred to the surroundings as work.

The membrane is now replaced by a frictionless piston on which the surroundings exert an external force. This force is assumed at all times to be in equilibrium with the force on the piston exerted by the gas in the cylinder (Figure 3.7(b)). The gas can now expand in a quasi-equilibrium process until it fills the container. This process

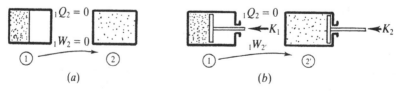

Fig 3.7 (a) Irreversible expansion. (b) Reversible expansion.

$1 \rightarrow 2'$ is reversible and energy is transferred to the surroundings in the form of reversible work. Given the same initial state, the final state of the gas will be different in the two situations.

In the uncontrolled and spontaneous irreversible process there is a loss of possible work, the so-called 'lost work', which, by way of fluid friction, has been dissipated into internal energy when the gas has reached equilibrium state 2 after the expansion. Note that the process in Figure 3.7(a)) would also proceed as described if the membrane were replaced by a frictionless piston that was suddenly moved all the way to the right. For this to be true, the movement of the piston should be so fast as compared to the expansion of the gas that the gas could never exert a pressure on the moving piston. Hence, no energy would be transferred as work to the surroundings.

Quasi-static stretching of an elastic wire, or a general three-dimensional deformation of an elastic body, follow form (20), and are examples of reversible forms of work, where σ is the stress (force per unit area), ε the strain (relative deformation) and $\mathscr{V}$ the volume of the system.

Stretching of a free liquid surface, i.e. production of new surface area, dA, requires the addition of energy in the form of work to the system containing the surface. The molecules close to the surface are influenced by the attractive forces of the molecules in the liquid phase in an asymmetric manner and are therefore in a state of higher energy than the molecules in the interior of the liquid. Macroscopically, this gives rise to a finite surface tension σ_s. This quantity can be interpreted as the force per unit length of a cut in the surface or as the energy per unit area of the free surface.

The lungs are an example of a biological system with a large surface area (about $100\,m^2$ in an adult subject). Here, the surface tension is of practical importance. During inspiration, the incoming air fills the branched system of canals (the bronchial tree) that ends with the near-spherical alveoli, across whose walls there is transport of oxygen and carbon dioxide. The single alveolus can be considered as an elastic membrane, covered by a thin film of liquid. The membrane is relatively compliant in comparison to the liquid film.

At equilibrium between the surface tension σ_s and the excess pressure Δp of the gas inside a spherical surface with radius R, the Laplace relation states that $\Delta p = 2\,\sigma_s/R$. If σ_s were to be constant, an increasing radius would lead to decreasing excess pressure and the alveoli would collapse when the excess pressure fell below a certain value. Furthermore, it would be impossible to distend alveoli of different radii with air from the same reservoir because the smaller alveoli would empty into the larger ones. Instabilities of this kind do not occur in normal lungs and experiments show that σ_s does not remain constant but increases with increasing radius in such a way that Δp always increases with increasing R. This property of σ_s is due to the presence of surface active matter (phospholipids) in the liquid film. When the surface of the film distends, the concentration of the surface active matter decreases in the interface between liquid and air and both σ_s and Δp increase. The process exhibits hysteresis, governed by the diffusion of matter between the free interface and the interior of the liquid.

The surface tension is about 0.06 N/m for water and about 0.05 N/m for blood plasma, but can vary between 0.002 and 0.04 N/m for the liquid films of alveoli when their total area changes by a factor of five (West, 1991).

Example 3.7

According to the data given above and for a maximal area of 100 m^2, the total surface energy $\sigma_s A$ varies between the values

$$0.04 \times 100 = 4\,J \quad \text{and} \quad 0.002 \times 20 = 0.04\,J. \tag{a}$$

The hysteresis over a cycle can be estimated to be 25% of the maximal energy so that the power due to surface tension at a respiration frequency of 12 per minute is estimated to be

$$0.25 \times 4 \times 12/60 = 0.2\,W. \tag{b}$$

The remaining contributions to respiratory power include frictional resistance of the flow of air and, to a lesser extent, hysteresis in the walls of the alveoli and lung tissue. The excess pressure to distend an alveolus can be calculated from the Laplace relation

$$\Delta p = 2\sigma_s/R. \tag{c}$$

The radii of the ≈ 300 million alveoli vary in the interval $R = 0.06 - 0.15$ mm, so that the excess pressure from surface tension can vary between

$$\Delta p = 2 \times 0.04/(0.15/1000) = 533\,Pa \tag{d}$$

and

$$\Delta p = 2 \times 0.002/(0.06/1000) = 66 \, \text{Pa}. \tag{e}$$

It should be noted that (c) can be rewritten through multiplication by $4\pi R^2 \, dR$ and integration as:

$$\int \Delta p \, d = \int \sigma_s \, dA, \tag{f}$$

which shows the equality between the work of the excess pressure and the increase in surface energy.

In living systems, muscles perform considerable *internal work* (pumping of blood, peristalsis of the gut, chewing of food, etc.) and *external work* (locomotion in water, on land or in air and during manual work). It should be noted that all forms of external work also involve internal muscular work to overcome friction and to accelerate and decelerate extremities.

3.5.3 Electric and magnetic work

Material systems can be influenced by electromagnetic fields in various ways. We have seen that the movement of an electric charge (a monopole) in an electric field from one potential to another potential involves work that can be conveniently interpreted as a change in potential energy. There is no corresponding magnetic effect because quantities of magnetization can only exist as dipoles (small magnets, each with a north pole and a south pole).

Electrically neutral dipole structures and magnetic dipole structures in a material system can be affected by an external electromagnetic field that varies over time. When the strength of the field increases, it will tend to rotate and align the dipoles according to the direction of the field. The degree of alignment is a measure of the polarization and magnetization, respectively, and both of these quantities are properties of state. The product of electrical (magnetic) field strength and the corresponding change in polarization (magnetization) (Table 3.2), is equal to the energy added to the material in the form of work necessary to overcome the resistance of the material to the alignment of dipoles.

If the strength of the field increases slowly in a loss-free material, the work will be reversible and the energy added as work can be recovered as work when the field is reduced. However, if the material is lossy, part of the work will be irreversible and cannot be recovered. This part of the added energy will be converted into internal energy. For a differential process, we have

$$\delta W_{em} = \delta W_{em}^{rev} + \delta W_{em}^{irr}. \tag{22}$$

For a periodically oscillating field, the irreversible contributions from hysteresis will be continuously converted into internal energy and the process appears as if the material were subject to an internal energy production. In an adiabatic system, this results in a local increase in temperature. The phenomenon is well known from the use of microwave ovens for thawing and heating of food. Here, the contribution to absorption and dissipation of energy comes primarily from the presence of water, whose dipole moment responds to electromagnetic waves in the frequency range 2–10 GHz. The radiation penetrates into the material containing water and the power is therefore dissipated relatively evenly throughout the material. This is particularly true for the lower frequencies that are characterized by longer wave lengths, in that frequency ν and wave length λ are related to the velocity of propagation of electromagnetic waves c, which is close to the velocity of light ($c = 3 \times 10^8$ m/s):

$$\nu\lambda = c. \tag{23}$$

All tissues in living systems, such as skin, fat, skeletal muscle and bone, react to electromagnetic radiation by absorption of energy. This fact is used in *diathermy* and *microwave therapy*. The absorption of the energy depends on the type of tissue and the applied frequency. A distinction is often made between low-frequency radiation (3–300 MHz) and high-frequency radiation (1–50 GHz) (Kresse, 1985). The resulting distribution of temperature in a tissue depends on the power dissipated, the heat capacity of the tissue and its ability to remove heat. Sources of radiation for microwave therapy typically have powers of between 25 and 250 W, depending on the area of the radiated field. These powers may be compared to a total heat production of normal human subjects at rest of some 60–80 W.

Atomic nuclei behave as magnetic dipoles. They can be influenced by externally imposed oscillating magnetic fields within a range of frequencies. When the frequency is the same as the natural resonance frequency of the nuclei, the latter emit signals and can therefore be identified. The procedure is called nuclear magnetic resonance NMR and can be used in NMR-tomography (see, e.g., Gillies, 1993) to obtain information on the chemical structure of organs. The power dissipated in tissue exposed to NMR is usually small.

Example 3.8

Induction of nuclear magnetic resonance to produce images in clinical situations is associated with maximum energy deposition of 1 W per kilogram tissue during periods lasting up to 20 minutes. The maximum total energy deposition is therefore $1 \times 20 \times 60$ kJ/kg. The quantity 1 W/kg is very similar to the heat production of normal human subjects at rest. Usually, however, exposures involve much smaller powers.

3.5.4   Chemical work

A particular type of energy exchange between systems may take place by way of so-called chemical coupling. Here, the energy released from a reaction in one system is received (and consumed) directly by a reaction in another system; the two reactions are said to be coupled. As will be shown later (in Section 4.5.5), the energy exchange can be expressed as

$$\delta W_{Ch} = \Sigma \; \mu_j \, dM_j, \tag{24}$$

where μ_j is the chemical potential (see (92)) representing the force, and dM_j is the change of mass, representing the corresponding displacement in 'chemical space' for component j of the reaction; δW_{Ch} represents energy in transit and, in contrast to heat, is useful energy, or free energy, and may be called (reversible) chemical work. As indicated in Figure 1.6, most of the energy consumption in biological systems involves energy conversion and energy transfer through coupled reactions (to be elaborated in Sections 4.5.5 and 8.4.2).

It should be noted that (24) has the same form as (2.2), representing a generalized expression for all forms of differential reversible work. However, it should only be included in the energy balance (in the term δW of the first law (2.5)) for a system which includes a chemical reaction that is coupled to another chemical reaction located outside the system.

3.6 Heat

The transfer of energy as heat between a system and its surroundings is closely related to the concepts of temperature and *diabatic* (heat conducting) contact between bodies, concepts that are implicit in the zeroth law of thermodynamics (2.4).

Experience shows that heat appears as energy in transit to a system when there are differences in the temperature between the system and its surroundings in diabatic contact. The nature of the contact can be material, which allows energy transfer by *conduction* (diffusion), or non-material, which allows energy transfer by *radiation* (waves, photon emission and absorption), or by a combination of both. Transfer of heat by conduction to the control surface from a fluid in motion is called *convection*. An *adiabatic* contact (perfect heat insulation) precludes the transfer of energy as heat.

The heat exchanged between a living system and its surroundings consists in part of a direct transfer by conduction, convection and radiation, which together are denoted *sensible heat*, and in part of an indirect transfer due to the evaporation or condensation of water, from and to the surface, denoted *evaporative heat*. The last-mentioned heat is of importance for homeothermic animals where evaporation takes place from the skin and in the respiratory organs (nose, mouth, bronchial tree and lungs). The evaporative heat loss in human subjects at rest and at room temperature is about 25% of the total heat loss, but it can amount to as much as 95% by way of sweating in warm and dry surroundings. The evaporative heat transfer is in fact a flow of enthalpy (see Example 4.8).

In the present terminology, heat energy also includes that part of radiation in the visible range of the electromagnetic spectrum that is absorbed by the chlorophyll molecules in green plants and transferred by coupling to synthesis of carbohydrate (see, however, the remarks in Section 5.4.3).

Also, heat Q is considered to be *positive* when it is *added* to the system. The energy transferred as heat is not a state property and depends on the nature of the process. Transfer of heat can be reversible or irreversible.

According to the second law (2.6), reversible addition of heat to a system can be expressed by a change in the state properties of the system, and for a differential reversible process, by

$$\delta Q_{rev} = T \, dS, \tag{25}$$

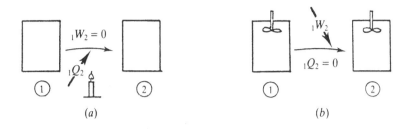

Fig 3.8 (a) Energy supply as heat. (b) Energy supply as work.

where the relation between T and S is given by the equation of state of the system and the process in question. It should be noted that the validity of (25) assumes that no irreversible process takes place. A real process in a given system approaches a reversible process when the energy added as heat takes place at vanishing temperature differences within the system and between the system and its surroundings. This phenomenon is discussed in more detail in Section 5.3.

Example 3.9

A gas is contained in a closed and rigid vessel at a given observed state 1. It can be brought to a new observed state 2 with a larger energy content by the supply of energy from the surroundings as heat (Figure 3.8(a)) or as work (Figure 3.8(b)). The observation is a confirmation of the equivalence of energy, irrespective of its form during supply. In addition, it is evident that when a given quantity of energy has been added to a system, it is impossible by observation of the increased energy to determine whether its origin was heat or work. It is therefore meaningless to talk about the heat content (or work content) of systems.

Inspection of (2.2) and (25) suggests that the reversible supply of energy as both heat and work could be described as a simple sum of products of generalized forces and associated displacement. However, microscopic and macroscopic differences make it desirable to make a distinction between the two forms of energy exchange.

As noted in Chapter 2, history makes it clear that four distinct concepts have caused epistemological difficulties, namely work, heat, absolute temperature and entropy. We can analyse the difference between heat and work from the microscopic point of view in the following way.

Microscopically, the state of a system is determined by the total internal energy U of the particle ensemble and by the distribution of this energy in possible quantum states. This distribution is determined by a number of boundary and volume conditions which can be assigned to a few macroscopically observable parameters such as the volume (of a fluid), geometrical dimensions (of a solid), surface area and surface curvature (of a crystal, liquid drop), degree of electrical polarization and magnetization, or the number of moles of components (of a mixture). These global (and extensive) parameters X_j determine, together with the internal energy U, the macroscopic state of the system, including its entropy, see (2.3).

A process involving the transfer of energy as heat is then characterized by a change in state of the system involving no changes in the parameters X_j mentioned above. On the microscopic level, this form of energy transfer is associated with the transport of photons (radiation) and with the transport of phonons (conduction) in material media. A process involving transfer of energy as work, on the other hand, is characterized by a change in state of the system that takes place solely as a result of changes in the parameters X_j. On the microscopic level, this form of energy transfer is associated with changes in the state of particles (e.g. polarization) or in the boundary conditions for the particle ensemble, such as the movement of an isolating wall (adiabatic change in volume).

For a more detailed treatment of microscopics, the reader is referred to texts on statistical thermodynamics, e.g. Giedt (1971), for an introduction, and Sonntag and van Wylen (1966) for a more complete description.

As pointed out in Example 3.9, a system at a given state, say at zero kinetic and potential energy, contains only internal energy. The magnitude of this energy is determined by the given state at equilibrium. The manner in which this state of energy has been achieved only depends on the preceding process in regard to how much energy was added or extracted, not the form of this energy.

Nevertheless, the internal energy of a system is sometimes characterized as thermal, elastic or chemical internal energy. Such concepts are based on the idea that the system, when undergoing a given process, will transfer energy to the surroundings in the form of heat, mechanical (elastic) work or chemical work (coupled reactions).

The relevance of such terminology is clear enough in some situations but not so in other situations. Suppose a vessel containing water is heated so that its internal (thermal) energy increases; the latter will subsequently decrease as a result of transfer of heat to the surroundings. A spring can be compressed so that its internal (elastic) energy increases and the latter can later decrease as a result of transfer of mechanical (elastic) deformation work on the surroundings. In a third example, we compress the air adiabatically in a bicycle pump so that the internal energy of the gas increases. In this case, however, we meet with difficulties in associating the increase in energy with either thermal or mechanical energy. In fact, in all three examples, the increase in internal energy, if any association is needed, should be associated with an increase in thermomechanical internal energy. In the first two examples, the mechanical and thermal energy, respectively, vanish, while the distinction is problematic in the third example. Therefore, we will speak of internal energy only. The change in internal energy in specific processes follows from the use of the first law (Chapter 4).

3.6.1   Temperature

As mentioned above, the transfer of energy as heat is closely associated with the concept of temperature. The zeroth law establishes the concept of equality of temperatures. The practical design of thermo-meters (based on thermal volume expansion, changes in electrical resistance, thermoelectric effects, optical pyrometry, etc.) establishes relative temperature scales. Reference points (triple point, melting points, etc.) establish the internal and mutual calibration of scales. A detailed description of the definition of the practical temperature scale and of reference points is given in Giedt (1971).

The absolute temperature is a thermodynamic state property implicit in the second and third laws of thermodynamics. Its zero reference point is absolute, while the unit of the scale (1 K = one degree kelvin) is arbitrary but fixed by international convention so that the triple point for water, prepared in a prescribed way, is 273.16 K = 0.01 °C. The thermo-dynamic definition of absolute temperature is given in Section 3.8.3 by (42).

3.6.2   Calculation of heat transfer

This subject is both extensive and complex. A comprehensive treatment is given in Lienhard (1981). Here we illustrate some of the basic principles for heat transfer by conduction, convection and radiation using simple stationary examples.

For one-dimensional heat *conduction*, the Fourier law

$$\dot{Q}/A = -k \, dT/dx \tag{26}$$

expresses the fact that the heat flux in the x-direction per unit time and area, $\dot{Q}/A$, is proportional to the magnitude of the temperature gradient

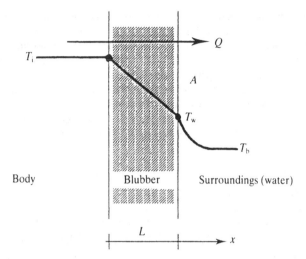

Fig 3.9 Heat transfer from a body through its insulating layer of blubber to the surroundings.

dT/dx and opposite to it in sign. The constant of proportionality k is the thermal conductivity. Under steady conditions, $\dot{Q}$ is constant and the temperature decreases linearly with x whenever k is a constant and $\dot{Q}$ is positive.

Heat transfer by *convection* from a surface with wall temperature T_w to a fluid (liquid or air) with bulk temperature T_b flowing past the surface can be described by Newton's law of cooling (Figure 3.9):

$$\dot{Q}/A = \bar{h}(T_w - T_b), \tag{27}$$

which defines the heat transfer coefficient $\bar{h}$. This can be determined empirically (Lienhard, 1981) or from theory (Arpaci and Larsen, 1984) and depends on the geometry, the properties of the fluid and the flow conditions.

The algebraic sign in (27) is determined so that the heat flux $\dot{Q}/A$ is positive in the direction of decreasing temperature.

Example 3.10

The blubber of seals has a low heat conductivity, about 0.1 W/m-K or five to six times lower than that of water. A 40 kg seal with an external surface area of $A = 1\,\mathrm{m}^2$ has a heat production of $\dot{Q} = 50\,\mathrm{W}$ and an insulating layer of blubber of thickness $L = 0.065\,\mathrm{m}$.

Calculate the external temperature T_w on the skin when the inner body temperature is $T_i = 37\,°C$ and the heat transport in the blood of the blubber can be neglected.

Integration of (26) over the blubber layer gives (Figure 3.9)

$$\dot{Q} = A\,k(T_i - T_w)/L, \qquad\qquad\qquad (a)$$

hence

$$T_w = T_i - \dot{Q}\,L/(A\,k) = 37 - 50 \times 0.065/(1 \times 0.1) = 4.5°\,C. \quad (b)$$

Example 3.11

Calculate the heat transfer coefficient for Example 3.10 when the temperature of the surrounding water (the free stream temperature) is $T_b = 0\ °C$. From (27),

$$\bar{h} = \dot{Q}/[A(T_w - T_b)] = 50/[1 \times (4.5 - 0)] = 11\ W/m^2/K. \qquad (a)$$

Example 3.12

In Examples 3.10 and 3.11, the same heat flux $\dot{Q}/A$ passes through two resistances in series, L/k in the blubber and $1/\bar{h}$ in the thermal boundary layer in the water near the skin.

Rewriting (a) in each example and adding the results, gives

$$(\dot{Q}/A)(L/k + 1/\bar{h}) = T_i - T_b \qquad\qquad\qquad (a)$$

$$(50/1)(0.65 + 0.09) = 37\,°C, \qquad\qquad\qquad (b)$$

which shows that heat flux multiplied by total resistance equals the total temperature difference (in analogy with Ohm's law for current, resistance and voltage). The large resistance, hence the large temperature drop, is in the blubber.

It should be noted that the total heat transfer $\dot{Q}$ from a living system is proportional to the surface area, cp. (26) and (27), for given temperature and heat transfer conditions. Since the length L is a measure of the linear size of the system, we have for the area $A \approx L^2$, the volume $V \approx L^3$, for the mass $= M \approx \rho\,L^3$, and for the heat transfer

$$\dot{Q} \approx A \approx M^{2/3}. \qquad\qquad\qquad (c)$$

Relation (c) could be expected to be valid for homeothermic animals of different sizes if they are geometrically similar, have the same density and similar conditions for heat transfer. In spite of these restrictions it is of interest to note that measurements on a whole range of such animals show the validity of the relation

$$\dot{Q} \approx M^{0.7}. \tag{d}$$

Heat transfer by *radiation* between material surfaces of different temperatures is due to thermal emission and absorption of photons. Thermal radiation can also be described as electromagnetic radiation. Because it is a thermal phenomenon, the energy transfer is regarded as heat, and not as work of the types described in Section 3.5.3. The exchange takes place with no hindrance in the case where there is a vacuum or a non-absorbing gas (such as atmospheric air) between the surfaces. Water transmits radiation in the visible range of the spectrum but absorbs the thermal radiation of longer wave lengths effectively in a layer of a few millimetres thickness. Radiation between surfaces is therefore of little importance for aquatic animals.

In the following, we consider thermal radiation, which in practice involves a continuous spectrum. This is in contrast to photosynthesis and photoelectric processes which involve band spectra, implying photon absorption at selected wave lengths or narrow bands of wave lengths.

The model of a perfect thermal radiator is called a *black body*. It absorbs all incoming radiation, and reflects none, so its absorptivity is $\alpha = 1$. The name has been chosen because our visual perception of a perfect absorber in the visible range of the spectrum is black. The black body emits radiation with the ideal Planck distribution over wave lengths, and the total emitted flux of energy per unit area is given by the *Stefan–Boltzmann law*:

$$E_0(T) = \sigma T^4, \tag{28}$$

where $\sigma = 5.67 \times 10^{-8} \text{ W/m}^2 - \text{K}^4$ is the Stefan–Boltzmann constant. The wave length at which the intensity in the Planck distribution is a maximum is inversely proportional to the absolute temperature of the body (Wien's law):

$$\lambda_{\max} = \text{constant}/T. \tag{29}$$

In the radiation from the sun ($T \approx 6000 \text{ K}$), this wave length is 540 nm; from a furnace (at 1000 K) it is about 3000 nm; and from a black body at

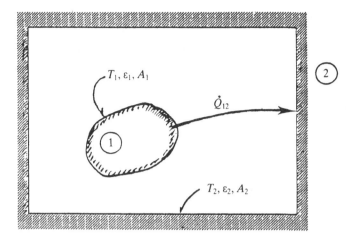

Fig 3.10 A grey isothermal body (1), enclosed in a room with grey isothermal walls (2).

room temperature (at 300 K) it is about 9000 nm. The radiation intensity at λ_{max} is proportional to T^5.

Many real surfaces can be treated as *grey bodies*, for which $\alpha < 1$, irrespective of wave length. At thermal equilibrium, $\varepsilon = \alpha$, where ε denotes the *emissivity*, and the total emitted energy flux (energy per unit area and time) is

$$E(T) = \varepsilon \, \sigma \, T^4. \tag{30}$$

The *reflectivity* of an opaque (non-transmitting) body is $\rho = 1 - \varepsilon$.

The calculation of heat transfer by radiation between surfaces generally requires an elaborate analysis involving integration (summation) over all contributions between all surface elements that can 'see' one another. Here we treat the simplest case of an isothermal grey body $(T_1, \varepsilon_1, A_1)$ contained in a room with grey and isothermal walls $(T_2, \varepsilon_2, A_2)$; see Figure 3.10. All surfaces are assumed to be opaque.

The calculation is simple when the body and walls are black. According to (28), the body emits the total energy $A_1 \, \sigma \, T_1^4$, which is absorbed by the walls, and it absorbs the energy $A_1 \, \sigma \, T_2^4$, which comes from the walls. Hence, the net transfer of energy from the body to the walls is

$$\dot{Q}_0 = A_1 \, \sigma(T_1^4 - T_2^4). \tag{31}$$

Next, for grey surfaces, the body only emits $A_1 \, \varepsilon_1 \, \sigma \, T_1^4$, of which the fraction $\rho_2 = 1 - \varepsilon_2$ is reflected from the walls. Of this, the body will absorb only the fraction ε_1, while the fraction $(1 - \varepsilon_1)$ is reflected back to the walls, and so on. The same considerations apply to the radiation from the walls, and the net result becomes

$$\dot{Q}_{12} = A_1 F_{12} \sigma(T_1^4 - T_2^4), \qquad (32)$$

where

$$F_{12} = 1/[1/\varepsilon_1 + (1/\varepsilon_2 - 1)A_{12}] \qquad (33)$$

is the *view factor*, seen to depend on geometry and surface properties. When the surfaces reflect diffusely, $A_{12} = 1$, and when they reflect specularly (like a mirror), $A_{12} = A_1/A_2$. Also, when $\varepsilon_1 > 0.8$ and $\varepsilon_2 > 0.8$, (33) can be approximated as

$$F_{12} = \varepsilon_1 \varepsilon_2, \qquad (34)$$

with an error of less than 5%. In the case where just one of the surfaces is a perfect reflector ($\varepsilon = 0$), we have $\dot{Q}_{12} = 0$.

Most surfaces of living organisms, and the surfaces which surround them under most circumstances, are characterized by a relatively large and diffuse emissivity ($\varepsilon \approx 0.9$) for radiation at prevailing temperatures. An illustrative example of differences here is the fact that snow has a low emisivity ($\varepsilon \approx 0.1$) for high-temperature radiation (it appears white to the eye) and a high emissivity ($\varepsilon \approx 0.9$) for radiation at body temperatures (it is almost black in the infrared spectral range).

Example 3.13

Calculate the heat transfer by radiation from a naked human being ($T_1 = 32\,°C = 305\,K$, $\varepsilon_1 = 0.9$, $A_1 = 1.8\,m^2$) situated in a room having cold walls ($T_2 = 7\,°C = 280\,K$, $\varepsilon_2 = 0.8$). From (32) and (34) we obtain

$$\dot{Q} = 1.8 \times 0.9 \times 0.8 \times 5.67 \times (3.05^4 - 2.80^4) = 185\,W. \qquad (a)$$

In addition, there is loss of heat due to convection to the air if its temperature is less than that of the body. Note that the heat loss (a) is more than twice that of the heat production at rest.

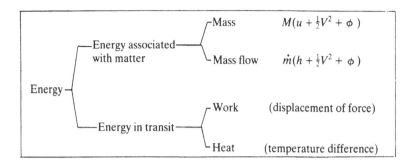

Fig 3.11 Summary of forms of energy.

3.7 Summary of energy forms

We summarize the classification and terminology related to energy used in this text in Figure 3.11.

The mass M of matter within the system possesses internal energy u, kinetic energy $\frac{1}{2}V^2$ and potential energy φ, all per unit of mass. The same is true for a mass flow $\dot{m}$, which enters or leaves the control volume, but in addition, it possesses a pressure–volume work associated with the flow. The sum of this work and the internal energy defines the enthalpy h ($\equiv u + pv$) so that the mass flow $\dot{m}$ possesses enthalpy, kinetic energy and potential energy.

The transfer of energy not associated with mass (i.e. energy in transit) is identified either as work or as heat. Work is identified with the displacement of a force (mechanical, electromagnetic) or, in the case of coupled chemical reactions, with chemical work. Heat is identified with the transfer of thermal energy in the direction of decreasing temperature with diabatic contact, as conduction, convection or radiation, and including radiation for photosynthesis.

3.8 State properties of a pure substance

The treatment in this section is limited to a one-component substance (pure matter) in a single phase, such as a gas, liquid or solid phase. Substances composed of several components (homogeneous mixtures) will be discussed in Section 3.10.

3.8.1 State

A material system can be described in terms of a number of characteristic properties, such as mass, geometric form, volume, surface area, electrical potential, electrical polarization, magnetization, chemical composition, pressure, stress, temperature, energy, entropy, colour, roughness, taste, smell, etc. In addition, there are a large number of molecular, atomic and nuclear properties. Of all these descriptive properties, some are relevant and some are irrelevant, all depending on the purpose of the description.

For the description of a system for the purpose of macroscopic thermodynamics analysis, the relevant properties are the *thermodynamic state properties*. Apart from kinetic and potential energy, the state properties are the internal energy of the system U and the macroscopic parameters X_j, changes in which can affect the internal energy as a result of the performance of reversible work.

Example 3.14

The volume of a gas in a cylinder is a state property which can change through displacement of a piston. The work performed here represents an exchange of energy between the gas and its surroundings, and the internal energy of the system is consequently changed.

On the other hand, the compressibility of a liquid is small and, in many cases, negligible. If the liquid is considered to be incompressible, its volume is constant. The volume does not determine the state of the system and it is not a state property. Also, the pressure of the liquid, which can be changed by an external force on the piston in a cylinder containing the liquid, is not a thermodynamic state property, but a mechanical or dynamic parameter.

The electric dipole moment of an incompressible substance, e.g. a piece of biological tissue, is a state property which can be changed by way of a change in the electric field strength in the substance. The work performed here is an exchange of energy between substance and surroundings and it can change the internal energy of the system.

The concept of a *thermodynamic state* implies the concept of an *equilibrium state*. A control mass which cannot exchange energy with its surroundings is said to be *isolated*. If all interactions between the parts of such a system are allowed to proceed, the system is said to have achieved a state of equilibrium when all processes have ceased. According to the *state postulate* (2.3), the entropy of a system exists at equilibrium and is a single-valued, continuous and differentiable function of the fundamental parameters of state:

$$S = S(U, X_j). \tag{2.3}$$

The state is completely determined by (2.3), which is the most general *equation of state*. As will be shown in Section 3.8.3, all other state properties can be derived from it.

3.8.2 Extensive and intensive state properties

The state properties of a material system can be divided into two types. These are the *extensive state properties*, such as the internal energy U, which are proportional to the mass of the system, and the *intensive state properties*, such as the temperature T, which are independent of the mass of the system. For example, if an ice cube is divided into two equal parts, the mass, the volume and the internal energy are halved, but the temperature remains unchanged.

State properties appear in a natural way in intensive–extensive pairs of associated (conjugated) quantities, as indicated in the expression for reversible work (2.2). This is also true when we consider extensive state properties per unit of mass, such as $u = U/M$, which are called *specific (extensive) state properties*. As an illustration, the state properties of a compressible medium are

Extensive:	U	S	$\mathcal{V}$
Specific extensive:	u	s	v
Intensive:		T	p

3.8.3 The Gibbs relation

This relation can be derived from the first and second laws of thermodynamics or from the postulate of state (2.3). The relation can be considered as a *compatibility condition* between state properties at equilibrium. It is of great importance in the formulation of alternative forms of equations of state and for the derivation of a number of general relations between properties of state.

Neglecting changes in kinetic and potential energy, and considering a reversible process in a control mass, the first law (2.5) and the second law (2.6) can be written as

$$dU = \delta Q_{rev} + \delta W_{rev}; \quad dS = \delta Q_{rev}/T. \tag{35}$$

The *Gibbs relation* is obtained after elimination of Q_{rev}, yielding

$$dU = T\,dS + \delta W_{rev}; \quad \delta W_{rev} = \Sigma F_j\,dX_j. \tag{36}$$

For a reversible process, the change in the sum of kinetic and potential energies can be shown to equal the work performed by external forces which does not contribute to δW_{rev}. Equation (36) is therefore valid in general.

Considering the *simple compressible substance*, for which the reversible work is

$$\delta W_{rev} = -p\,d\mathscr{V}, \tag{37}$$

the Gibbs relation is

$$dU = T\,dS - p\,d\mathscr{V}, \tag{38}$$

or, expressed in terms of specific state properties,

$$du = T\,ds - p\,dv. \tag{39}$$

Relevant forms for other substances are obtained by substituting expressions for δW_{rev} from Table 3.2 in (36).

Differentiating the state postulate (2.3) for the compressible medium, i.e. $S = S(U,\mathscr{V})$, or the inverse function,

$$U = U(S,\mathscr{V}), \tag{40}$$

we obtain

$$dU = (\partial U/\partial S)_{\mathscr{V}}\,dS + (\partial U/\partial\mathscr{V})_S\,d\mathscr{V}. \tag{41}$$

Comparing (38) with (41) reveals the definitions of thermodynamic temperature and pressure:

$$T = (\partial U/\partial S)_{\mathscr{V}}; \quad p = -(\partial U/\partial\mathscr{V})_S. \tag{42}$$

In addition, it can be seen that the state properties U, S, $\mathscr{V}$, T, p are not independent but, rather, are constrained by the Gibbs relation.

3.8.4 Equations of state in terms of measurable quantities

For a given compressible substance, the fundamental relation (40) is a general equation of state which, if known, would describe the relation between all the state properties of the substance. Internal energy U and entropy S cannot be measured directly, but changes in these quantities can be determined indirectly through measurable quantities such as p–v–T data and heat capacities c_v and c_p.

Thus, differentiating the function $u = u(T,v)$ and using the Gibbs relation (39), and differential relations derived from this (see e.g. van Wylen and Sonntag, 1965), we may derive the following relations in terms of specific state properties:

$$du = c_v \, dT + [T(\partial p/\partial T)_v - p] \, dv, \tag{43}$$

$$ds = (c_v/T) \, dT + (\partial p/\partial T)_v \, dv, \tag{44}$$

where

$$c_v \equiv (\partial u/\partial T)_v \tag{45}$$

defines the *specific heat at constant volume*, which can be measured by calorimetry. Since the relation $p = p(T,v)$ can also be measured, we can calculate u and s at an arbitrary state (T,v) by integration of (43) and (44) from a reference state (u_0, s_0) at (T_0, v_0).

Furthermore, in connection with the derivation of the first law for a control volume (Section 4.2) it is convenient to define the *derived* state property

$$\text{enthalpy: } H = U + p\mathcal{V}; \ h = u + p \, v. \tag{46}$$

Differentiating the function $h = h(T,p)$ and using the Gibbs relation, we may derive the relations

$$dh = c_p \, dT - [T(\partial v/\partial T)_p - v] \, dp, \tag{47}$$

$$ds = (c_p/T) \, dT - (\partial v/\partial T)_p \, dp, \tag{48}$$

where

$$c_p \equiv (\partial h/\partial T)_p \tag{49}$$

defines the *specific heat at constant pressure*. Given $c_p(T,p)$ and $p = p(T,v)$, changes in enthalpy h and entropy s can be calculated from (47) and (48) by integration from one state to another state.

Examples of equations of state for the simple cases of an ideal gas and an ideal (incompressible) liquid are given in the following section.

3.8.5 The ideal gas

By definition, an ideal gas follows the equation of state

$$p\mathcal{V} = MRT \text{ or } p\mathcal{V} = N\hat{R}T, \tag{50}$$

or, after division by mass M and mole number N, respectively:

$$pv = RT \text{ or } p\hat{v} = \hat{R}T, \tag{51}$$

where $v = \mathcal{V}/M$ is the specific volume and $\hat{v} = \mathcal{V}/N$ the molar volume, R is the *gas constant* (on a mass basis), $\hat{R}$ is the *universal gas constant* (on a molar basis), and these constants are related by the definitions

$$M = \hat{M}N; \quad \hat{R} = \hat{M}R; \quad \hat{R} = 8.314 \text{ J/mol-K}$$
$$= 0.082\ 07 \text{ atm-l/mol-K}, \tag{52}$$

where $\hat{M}$ is the *molar mass* (molecular weight). The symbol (ˆ) is used systematically in this text for *molar* quantities in order to distinguish these from the corresponding specific quantities per unit of mass.

Example 3.15

The gas constant for dry air, whose molar mass is 28.97 g/mol (see Example 3.19), is calculated as

$$R \text{ (air)} = 8.314/28.97 = 0.287 \text{ J/g-K.} \tag{a}$$

The density $\rho = 1/v$ of air at atmospheric pressure 1.013 bar $= 1.013 \times 10^5 \text{ N/m}^2$ and room temperature 20 °C $= 273.15 + 20 \approx 293$ K is calculated from (51) as

$$\rho = p/(RT) = 1.013 \times 10^5/(287 \times 293.15) = 1.204 \text{ kg/m}^3. \tag{b}$$

Accordingly, 10 l of air at this state has the mass

$$M = \rho\mathcal{V} = 1.204 \times 0.01 = 0.012\ 04 \text{ kg} = 12.04 \text{ g.} \tag{c}$$

The amount, or the quantity, of a gas is often expressed as $\mathcal{V}_{STP}$ (the volume at the standard state $T = 0$ °C, $p = 1.0$ atm), or, more precisely, as $\mathcal{V}_{STPD}$ (the volume at the standard state of dry gas, i.e. of the gas remaining after the water vapour has been removed). For 10 l of dry air

$$\mathcal{V}_{STP} = \mathcal{V}(T_S/T)(p/p_S) = 9.32 \text{ l.} \tag{d}$$

Example 3.16

Show that one mole of an ideal gas takes up 22.4 l at 0 °C and 760 mm Hg = 1 atm = 1.013 bar. From (51):

$$\hat{v} = \hat{R}T/p = 8.314 \times 273/1.013 \times 10^5 = 0.0224 \text{ m}^3 = 22.4 \text{ l/mol}.$$

It follows that $\mathcal{V}_{STP} = 22.4$ l for one mole of any gas that satisfies (50).

The simple p–v–T relation (51) implies considerable simplification of the equation of state expressed in terms of measurable quantities. Thus, differentiation of (51) gives

$$T(\partial p/\partial T)_v = T(R/v) = p \text{ and } T(\partial v/\partial T)_p = T(R/p) = v,$$

so that the last terms in (43) and (47) disappear, giving

$$du = c_v \, dT, \tag{53}$$

$$dh = c_p \, dT. \tag{54}$$

This result implies that u, h, c_v and c_p *for an ideal gas depend solely on the absolute temperature*. In addition,

$$c_p - c_v = R \text{ or } \hat{c}_p - \hat{c}_v = \hat{R}. \tag{55}$$

For small changes in temperature, c_p and c_v can be considered to be constant and integration of (53) and (54) gives

$$u - u_0 = c_v \, (T - T_0); \quad h - h_0 = c_p(T - T_0). \tag{56}$$

Similarly, integration of (44) and (48) gives

$$s - s_0 = c_v \ln (T/T_0) + R \ln (v/v_0), \tag{57}$$

or

$$s - s_0 = c_p \ln (T/T_0) - R \ln (p/p_0). \tag{58}$$

It follows that changes in internal energy, enthalpy and entropy between two states (T,p,v) and (T_0,p_0,v_0) can be readily calculated.

3.8.6 The incompressible liquid

Liquids can usually be considered as being incompressible for small changes in pressure and temperature:

$$dv = 0; \quad v = 1/\rho = \text{constant}, \tag{59}$$

so that the volume is not a state property. The fundamental relation (40) takes the form $U = U(S)$, and the Gibbs relation reduces to $dU = T\,dS$. Expressed in terms of measurable quantities, changes in u and s depend solely on the temperature:

$$du = c\,dT; \quad u - u_0 = c(T - T_0), \tag{60}$$

$$ds = (c/T)\,dT; \quad s - s_0 = c\ln(T/T_0), \tag{61}$$

where $c = dU/dT$ is the (only) specific heat, here assumed to be constant. According to the definition, $h = u + pv$, the enthalpy is pressure-dependent so that

$$dh = c\,dT + v\,dp; \quad h - h_0 = c(T - T_0) + v(p - p_0). \tag{62}$$

When a liquid is heated, the first term in (62) dominates, while an isothermal pumping process involves only the last term.

In subsequent applications, we will treat liquids and tissues in living systems as incompressible liquids.

Example 3.17

Calculate the change in internal energy of a cold beer (0.33 l at 7 °C) when it is heated to body temperature (37 °C) after ingestion. Multiplying (60) by the mass and using $c = 4.18$ kJ/kg-K for water, gives

$$U_2 - U_1 = M(u_2 - u_1) = 0.33 \times 4.18 \times (37 - 7) = 41.4\,\text{kJ}. \tag{a}$$

This energy change corresponds to the heat production at rest of a normal human subject (about 70 W) during a period of ten minutes.

Example 3.18

Calculate the change in enthalpy of blood when subjected to an isothermal increase in pressure of 16 kPa (120 mm Hg). From the second term in (62), we have

$$h_2 - h_1 = v(p_2 - p_1) = 10^{-3} \times 16\,000 = 16\,\text{J/kg}. \tag{a}$$

At a pulse rate at rest of 70 beats per minute and a stroke volume of 70 ml, the power is

$$\dot{m}(h_2 - h_1) = 0.070 \times 1 \times (70/60) \times 16 = 1.3 \, \text{W}. \tag{b}$$

This is about 1000 times the power associated with the kinetic energy of the flow (cp. Example 3.3).

3.9 Mixtures

The term 'pure substance' implies a homogeneous medium of constant chemical composition, e.g. an element (N_2, O_2, C, etc.) or a stable chemical component (H_2O, NH_3, $C_6H_{12}O_6$, etc.).

A *non-reactive mixture* is a homogeneous mixture of several components, each of which has a constant chemical composition. Humid air is a non-reactive mixture of, mainly, N_2, O_2 and H_2O, and a mixture of alcohol and water is a non-reactive mixture of H_2O and C_2H_5OH. The components of such mixtures are assumed not to react chemically to produce other components, but they will usually be non-ideal in the sense that they interact so that specific extensive state properties of a component within and outside the mixture (at a given pressure and temperature) will differ. The composition of a mixture can be changed by the addition or removal of components or by a change of phase (humid air). In many situations, a non-reactive mixture (e.g. dry air) can be treated as pure matter. This is practical and permissible as long as its composition remains constant.

A *reactive mixture* is a mixture of several components, of which two or more are involved in one or several chemical reactions (oxidation, reduction, etc.) The composition of the mixture will be changed by chemical reaction or by the addition or removal of components.

A *homogeneous* chemical reaction proceeds uniformly and simultaneously in all volume elements of a system composed of a one-phase homogeneous mixture (e.g. combustion of gaseous fuel mixed with air). A *heterogeneous* chemical reaction proceeds in space along surfaces between two or more phases (combustion of liquid or solid fuel, biochemical reactions at the surfaces of internal structures of cells). A catalysed reaction proceeds in the presence of catalysts which change the rate of the reaction while remaining chemically unchanged.

The mathematical formulation for a thermodynamic system that involves a mixture has been shown in Figure 2.1. Conservation of mass must be satisfied for each component. Any chemical reactions in the system must be specified by appropriate reaction equations, each of which must be formulated so that the mass of each component is conserved. Reaction rates may be a part of the solution of the problem or they may be specified (by constitutive relations). When chemical equilibrium can be assumed for a given reaction, the associated equilibrium conditions must be specified.

3.9.1 Composition

The composition of a mixture can be specified by the number of molecules of each component or by the number of moles N_j, $j = 1,2,\ldots,k$. This number is defined as the number of molecules divided by Avogadro's number, 6.022×10^{23} molecules per mole. One mole of a component is the number of grams specified by the molar mass $\hat{M}_j$.

Each extensive state property has a corresponding specific extensive property. The specific composition of a homogeneous k-component mixture is thus determined quantitatively by the *mole fractions* of the components of the mixture:

$$\chi_j = N_j/N; \quad \Sigma N_j = N; \quad \Sigma \chi_j = 1, \tag{63}$$

where the summation here and in the following is from 1 to k.

Corresponding to the definition of molar mass (molecular weight) of component j:

$$\hat{M}_j = M_j/N_j, \tag{64}$$

we can define the molar mass of the mixture

$$\hat{M}_m = M/N, \tag{65}$$

which can conveniently be expressed in terms of the mole fraction of the components

$$\hat{M}_m = \Sigma \hat{M}_j N_j/N = \Sigma \chi_j \hat{M}_j. \tag{66}$$

Example 3.19

Apart from negligible fractions of argon, carbon dioxide etc., dry atmospheric air consists of 21 mole% oxygen (O_2) and 79 mole% nitrogen (N_2), hence

$$\chi(O_2) = 0.21; \quad \chi(N_2) = 0.79. \tag{a}$$

Since $\hat{M}(O_2) = 32.00$ g/mol and $\hat{M}(N_2) = 28.02$ g/mol, the molar mass of dry air is calculated from (66) as

$$\hat{M}_{air} = 0.21 \times 32.00 + 0.79 \times 28.02 = 28.97 \text{ g/mol}, \qquad (b)$$

which for many applications can be approximated to 29.0 g/mol.

The composition of liquid mixtures can be specified by concentrations of dissolved matter (*solute*) in the liquid in which it is dissolved (*solvent*), e.g. for component j by its *molarity* (number of moles per litre of mixture):

$$c_j = [B_j] = N_j/\mathscr{V}, \qquad (67)$$

where B is the chemical symbol of the component. The unit mol/l is denoted by M (*molar*). Alternatively, the *mass concentration* can be used, i.e.

$$\rho_j = M_j/\mathscr{V} = \hat{M}_j c_j. \qquad (68)$$

For dilute liquid mixtures, the density of solvent and mixture can be assumed to be the same so that calculation in terms of mole fraction is facilitated. Other measures include, for instance, for a binary solution of solute 1 in solvent 2, the *molality* of the solute N_1/M_2 (number of moles per unit mass of solvent), and the *mass fraction* $M_1/(M_1 + M_2)$.

The compositions of mixtures of gases are often given in terms of the *partial pressure* p_j, or the partial volume $\mathscr{V}_j$ of the component. For ideal gases (Section 3.10.4)

$$p_j = \chi_j p; \quad \mathscr{V}_j = \chi_j \mathscr{V}, \qquad (69)$$

where p and $\mathscr{V}$ are the total pressure and volume, respectively, of the mixture.

Example 3.20

The molarity of a physiological salt solution is $c(NaCl) = 0.154$ $M(NaCl)$ in water, made up by adding to 1 l of water

$$M(NaCl) = c(NaCl) \times \mathscr{V} \times \hat{M}(NaCl) = 0.154 \times 1 \times 58.44$$
$$= 9 \text{ g NaCl}. \qquad (a)$$

The mass concentration is 9 g/l.

NaCl is practically completely dissociated in this solution, which therefore consists of three components, Na^+, Cl^- and H_2O, having mole fractions

$$\chi(Na^+) = \chi(Cl^-) = (9/58.45)/(9/58.45 + 1000/18.02)$$
$$= 0.0028, \tag{b}$$

$$\chi(H_2O) = 1 - 2 \times 0.0028 = 0.9972. \tag{c}$$

Given the degree of dissociation for a mixture, the mole fractions of solvent, solute and its dissociated components can be calculated in a similar way.

3.9.2   Conservation of mass

For the case of a mixture, the conservation of mass for a control volume can still be written in the form of (9):

$$dM/dt + \Sigma \dot{m}_{out} - \Sigma \dot{m}_{in} = 0, \tag{70}$$

provided mass and mass flows include contributions from all components:

$$M = \Sigma M_j, \quad \dot{m} = \Sigma \dot{m}_j. \tag{71}$$

However, a complete formulation requires a mass balance for each species. For a mixture of k components, this is achieved by writing a total of k mass balance equations, either one for each component or one for each of k-1 components plus (70) for the mixture. If chemical reactions take place, the mass (and mole number) of the reacting species is not conserved but can increase or decrease.

Denoting by $\dot{M}_{ij}$ the mass production (negative in the case of mass consumption) of the jth component in the ith of r chemical reactions in the system, the *mass balance* for the jth component is

$$dM_j/dt + \Sigma_{out} \dot{m}_j - \Sigma_{in} \dot{m}_j = \Sigma_r \dot{M}_{ij}. \tag{72}$$

We note that (72) has the general form of (6), where the source term on the right-hand side is the total production per unit of time of component j in the system. In non-uniformly distributed systems, mass diffusion through the control surface must be included as an additional term, including the flux $\dot{\Phi}_j$ on the right-hand side of (72) (see Section 7.2.2). Summation of (72) for all components $j = 1,2,. . .,k$ yields the total mass conservation (70), implying that $\Sigma_j \Sigma_{ij} \dot{M}_{ij} = 0$ and $\Sigma_j \dot{\Phi}_j = 0$.

Reactive mixtures are conveniently described in terms of mole numbers and (72) can be written as a *mole balance* for component j:

$$dN_j/dt + \Sigma_{out}\, \dot{n}_j - \Sigma_{in}\, \dot{n}_j = \Sigma_r\, \dot{N}_{ij}, \tag{73}$$

where the source term on the right-hand side is the total production per unit of time of component j for the r reactions in the system. Summation of (73) for all components yields the *total balance* for the system:

$$dN/dt + \Sigma_{out}\, \dot{n} - \Sigma_{in}\, \dot{n} = \Sigma_j \Sigma_r\, \dot{N}_{ij}. \tag{74}$$

Mole numbers are not usually conserved in chemical reactions, making the right-hand side of (74) different from zero, so (74) is of less interest and mass conservation is conveniently stated by k equations (73).

3.9.3 Reaction equations

An important element in the description of a chemically reactive system is a specification of the chemical transformations that take place. This is done for each reaction by a *reaction equation*. For example

$$\underset{\text{reactants}}{C_6H_{12}O_6 \text{ (s)} + 6O_2 \text{ (g)}} \quad \rightarrow \quad \underset{\text{products}}{6CO_2 \text{ (g)} + 6H_2O \text{ (}l\text{)}} \tag{75}$$

expresses the net result of a number of intermediate reactions which characterize combustion of glucose (solid $= s$, or solution in water $= aq$) with oxygen (gas $= g$) to produce carbon dioxide (g) and water (liquid $= l$). This equation is written in terms of moles and, when properly balanced, satisfies the requirement of mass conservation through conservation of *atom numbers* for each of the elements. In terms of mass, e.g. for 1 mole of glucose $= 180$ g $C_6H_{12}O_6$, we obtain

$$\begin{aligned} &180 \text{ g } C_6H_{12}O_6 \text{ (s)} + 192 \text{ g } O_2 \text{ (g)} \\ &\rightarrow 264 \text{ g } CO_2 \text{ (g)} + 108 \text{ g } H_2O \text{ (l)} \end{aligned} \tag{76}$$

where the mole masses have been taken to their integer values.

Usually, reaction equations are written on a mole basis. If B_j denotes the chemical symbol for components $j = 1,2,\ldots,k$, and ν_j the stoichiometric coefficients, we can use the compact form

$$0 \leftrightarrows \Sigma_1^k \nu_j B_j; \; \nu_j \begin{cases} < 0, \text{ reactant} \\ = \quad 0, \text{ non-participant} \\ > 0, \text{ product} \end{cases} \tag{77}$$

The unit for ν_j depends on the choice of mole number for one of the components, but it is typically the mole of component j per mole of one of the reactants in (75), such as glucose.

The stoichiometric coefficients in the balanced reaction equation ensure conservation of mass, and for a single reaction the source term in (74) can be written as

$$\dot{N}_j = \nu_j \, \xi, \tag{78}$$

where ξ is the *molar rate of reaction*, typically with respect to one of the reactants (in ((75)) moles of glucose per second). The sign of ν_j ensures, according to (77), that the source term in (78) is negative for reactants and positive for products.

For r simultaneous reactions $i = 1,2,\ldots,r$ with stoichiometric coefficients ν_{ij}, involving k components:

$$0 \to \Sigma_j \, \nu_{ij} \, B_j, \quad i = 1,2,\ldots,r, \tag{79}$$

the source term in (74) becomes

$$\dot{N}_j = \Sigma_r \, \dot{N}_{ij} = \Sigma_r \, \nu_{ij} \, \xi_i, \tag{80}$$

where ξ_i is the reaction rate of the ith reaction.

3.10   State properties of mixtures

Living systems are made up of gaseous, liquid and solid mixtures. Thermodynamic analysis of such systems require knowledge of the state properties of mixtures. Since the state properties of each component in pure form are often known it would be desirable to be able to calculate from these the state properties of the mixtures. This, however, turns out to be a difficult task because the individual components usually interact at the molecular level when they are present in a mixture. In this way they change the 'private' properties of each other and we have a *non-ideal mixture*, i.e. a mixture of dependent components. If there are no interactions, we speak of an *ideal mixture*, i.e. a mixture of independent components. Clearly, if two identical components were mixed, we would have an ideal mixture. Gases that obey the ideal gas law make up ideal mixtures. Dilute mixtures of liquids (solutions) can often be treated as being ideal since they follow the laws of Raoult and Henry for both solvent and solute. In general, however, mixtures of liquid and solid matter are non-ideal.

Apart from the first two sections, which deal with the fundamental theory, the formal treatment is limited to ideal mixtures. Introductions to non-ideal mixtures are found in Prigogine and Defay (1954).

3.10.1 Relations between state properties of mixtures

According to the concept of state, the thermodynamic equilibrium state for a homogeneous compressible mixture is uniquely determined by the internal energy U, the entropy S, the volume $\mathscr{V}$ and the mole numbers of the components N_1, N_2,. . . (or their masses M_1, M_2,. . .) (Figure 3.12). This is expressed by the *fundamental* relation

$$U = U(S,\mathscr{V}, N_1, N_2,. . .). \tag{81}$$

The differential of (81):

$$dU = \left(\frac{\partial U}{\partial S}\right)_{\mathscr{V},N_i} dS + \left(\frac{\partial U}{\partial \mathscr{V}}\right)_{S,N_i} d\mathscr{V} + \Sigma \left(\frac{\partial U}{\partial N_j}\right)_{S,\mathscr{V},N_{i\neq j}} dN_j \tag{82}$$

is the *Gibbs relation*

$$dU = T\,dS - p\,d\mathscr{V} + \Sigma\hat{\mu}_j\,dN_j. \tag{83}$$

Comparison with (82) gives the formal definitions of intensive state properties of absolute temperature T, pressure p and chemical potential $\hat{\mu}_j$ of component $j = 1,2,. . .,k$. The Gibbs relation (83) expresses the fact that the energy of the system can change as a result of changes in entropy, voiume or mole number. The last-mentioned change of the composition of the mixture can result from the addition or removal of a number of moles of one or several components, either through exchange with the surroundings or through chemical reactions within the system.

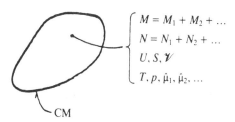

Fig 3.12 A homogeneous compressible mixture.

The *Euler form* corresponding to (81) and (83) is

$$U = T\,S - p\mathscr{V} + \Sigma\hat{\mu}_j\,N_j. \tag{84}$$

This relation can be derived by dividing the control mass in Figure 3.12 into differential elements. For each of these elements (81) is valid and dU and other differentials in (83) now refer to the quantities of extensive state properties in the differential element. If the system is homogeneous (intensive state properties T, p and $\hat{\mu}_j$ are the same everywhere), the contribution of all elements can be summed to give $\Sigma dU = U$, etc., and thereby (84). When (84) is differentiated and (83) is subtracted, we obtain the *Gibbs–Duhem relation*

$$0 = S\,dT - \mathscr{V}\,dp + \Sigma N_j\,d\hat{\mu}_j, \tag{85}$$

which expresses the relation between differential changes in intensive state properties for processes that proceed through equilibrium states.

The foregoing relations can also be expressed in terms of the supplementary state properties, introduced as definitions:

$$\left.\begin{array}{ll} H = U + p\mathscr{V} & \text{(enthalpy)} \\ A = U - TS & \text{(Helmoltz function)} \\ G = H - TS = U + p\mathscr{V} - TS & \text{(Gibbs function)} \end{array}\right\}. \tag{86}$$

Here, the enthalpy is a convenient function in the first law for analysis of a control volume (Section 4.2); the Helmholtz and, particularly, the Gibbs functions are convenient functions for the treatment of chemical mixtures and reactions, and for exergy analyses (Chapter 8).

3.10.2 Partial specific state properties

Comparison of the definition of the Gibbs function (86) and the Euler form (84) yields

$$G = \Sigma\hat{\mu}_j\,N_j. \tag{87}$$

The differential of (87) is

$$dG = \Sigma N_j\,d\hat{\mu}_j + \Sigma\,\hat{\mu}_j\,dN_j \tag{88}$$

or, after elimination of the first term on the right-hand side by use of (85):

$$dG = -S\,dT + \mathscr{V}\,dp + \Sigma\hat{\mu}_j\,dN_j. \tag{89}$$

This result, which is the Gibbs relation expressed in terms of the Gibbs function, could also have been obtained by introducing (83) in the differential of (86).

Because of the unique functional relation between independent state properties, we now perceive G to be a function of $T, p, N_1, N_2, \ldots$:

$$G = G(T, p, N_1, N_2, \ldots). \tag{90}$$

The differential of (90):

$$dG = (\partial G/\partial T)_{p,N_i} dT + (\partial G/\partial p)_{T,N_i} dp + \Sigma (\partial G/\partial N_j)_{T,p,N_{i\neq j}} dN_j, \tag{91}$$

shows, by comparison with (89), how extensive state properties S and $\mathscr{V}$ can be expressed as partial derivatives of G and – of particular importance – that the *chemical potential* μ_j of component j in the mixture is the partial specific Gibbs function for the component

$$\hat{\mu}_j(T,p,N_1,N_2, \ldots) = (\partial G/\partial N_j)_{T,p,N_{i\neq j}}. \tag{92}$$

This state property is of central importance in chemical thermodynamics. Being a specific state property, independent of the total mole number of the system, we can also write

$$\hat{\mu}_j = \hat{\mu}_j(T, p, \chi_1, \chi_2, \ldots, \chi_{k-1}), \tag{93}$$

where the specific composition of the mixture is uniquely determined by $k - 1$ mole fractions, cp. (63).

Example 3.21

Show that the chemical potential ((92)) can also be expressed by

$$\hat{\mu}_j = (\partial A/\partial N_j)_{T,\mathscr{V},N_{i\neq j}}. \tag{a}$$

The Gibbs relation expressed in terms of the Helmholtz function is obtained by the differential of A from (86) and the use of (83):

$$dA = -S \, dT - p \, d\mathscr{V} + \Sigma \hat{\mu}_j \, dN_j. \tag{b}$$

If A is perceived as being a function of $T, \mathscr{V}$ and N_j, which are the natural variables for A, inspection of (b) immediately gives (a).

The function A is convenient when T and $\mathscr{V}$ are constant.

It should be noted that expression (87) can be derived from (89) (cp. (91)) through integration at constant T, p, χ_i. In the same manner, for other extensive state properties perceived to be functions of T,p,N_1, $N_2,\ldots$, we can derive the relations

$$\left.\begin{array}{ll} \mathscr{V} = \Sigma \hat{\bar{v}}_j N_j, & U = \Sigma \hat{\bar{u}}_j N_j \\[2mm] H = \Sigma \hat{\bar{h}}_j N_j, & S = \Sigma \hat{\bar{s}}_j N_j \end{array}\right\}, \quad (94)$$

where partial molar state properties, in analogy with (92), are defined by

$$\left.\begin{array}{ll} \hat{\bar{v}}_j = (\partial \mathscr{V}/\partial N_j)_{T,p,N_{i\neq j}}, & \hat{\bar{u}}_j = (\partial U/\partial N_j)_{T,p,N_{i\neq j}} \\[2mm] \hat{\bar{h}}_j = (\partial H/\partial N_j)_{T,p,N_{i\neq j}}, & \hat{\bar{s}}_j = (\partial S/\partial N_j)_{T,p,N_{i\neq j}} \end{array}\right\}. \quad (95)$$

The properties are also considered to be functions of T, p, χ_1, $\chi_2,\ldots,\chi_{k-1}$. For the chemical potential, we have, in accordance with (86) and (92),

$$\hat{\mu}_j = \hat{\bar{h}}_j - T\hat{\bar{s}}_j. \quad (96)$$

Similar relations corresponding to other definitions in (86) apply to partial molar enthalpy and Helmholtz function for each component in the mixture.

It should be noted that for a 'mixture' consisting of one component only, i.e. pure matter, we obtain (94) and (95) that $\mathscr{V} = \hat{\bar{v}}_1 N_1$, and $\hat{\mathscr{V}}_1 = \partial \mathscr{V}/\partial N_1 = \mathscr{V}/N_1 = \hat{v}_1$, where $\hat{v}_1$ is the molar volume of component 1 as pure matter, a 'private' state property. Analogously, (87) and (92) give $\hat{\mu}_1 = \hat{g}_1 = G/N_1$, where the molar Gibbs function of pure matter is obtained, according to (86), as

$$\hat{g}_1 = \hat{h}_1 - T\hat{s}_1 \quad \text{(pure matter)}. \quad (97)$$

We follow common practice and use the notation $\hat{g}_j$ for the chemical potential of a pure substance.

3.10.3 Ideal mixtures

By definition, the change in volume and change in enthalpy are both zero for components that form an ideal mixture at constant temperature and pressure. Hence, there is no change in total volume and there is no 'heat of mixing'. It follows that

$$\hat{\bar{v}}_j = \hat{v}_j, \quad \hat{\bar{u}}_j = \hat{u}_j, \quad \hat{\bar{h}}_j = \hat{h}_j \quad \text{(ideal mixture)}, \quad (98)$$

so that the 'private' state properties can be used in (94) for calculation of the total volume, internal energy and enthalpy of the mixture. Since any process of mixing is irreversible, the entropy, however, will increase and, thereby, according to (96), decrease the Gibbs function:

$$\hat{\bar{s}}_j - \hat{s}_j > 0, \quad \hat{\mu}_j - \hat{g}_j = -T(\hat{\bar{s}}_j - \hat{s}_j) < 0. \tag{99}$$

The change in entropy on mixing can be calculated by integration of (44) for component j from the state in pure form to the state in the mixture, characterized by the partial volume $\mathscr{V}_j$ and the volume $\mathscr{V}$ respectively. For constant (T, p), we have

$$\hat{\bar{s}}_j - \hat{s}_j = \int_{\mathscr{V}_j}^{\mathscr{V}} (\partial p/\partial T)_v \, dv. \tag{100}$$

For an ideal gas, (51) gives $(\partial p/\partial T)_v = \hat{R}/\hat{v}$. Substituting this into (100) and using (69), we obtain

$$\hat{\bar{s}}_j - \hat{s}_j = -\hat{R} \ln \chi_j, \tag{101}$$

which is positive since $\chi_j < 1$.

The concept of ideality implies that the components are indepenent in the thermodynamic sense. Therefore, result (101) is valid for any ideal mixture, gas, liquid or solid matter.

Use of (99) and (101) gives, directly,

$$\hat{\mu}_j = \hat{g}_j + \hat{R}T \ln \chi_j, \tag{102}$$

where $\hat{g}_j = \hat{h}_j - T\hat{s}_j$ is calculated from tables or state equations for the component as pure matter at (T, p).

When k components are mixed to produce an ideal mixture at (T, p), the following changes in state properties are obtained:

$$\left. \begin{array}{l} \Delta \mathscr{V}_m = 0, \quad \Delta U_m = 0, \quad \Delta H_m = 0 \\ \Delta S_m = -\hat{R} \sum_1^k N_j \ln \chi_j, \quad \Delta G_m = \hat{R}T \sum_1^k N_j \ln \chi_j \end{array} \right\}, \tag{103}$$

where m denotes mixture.

3.10.4   Ideal gas mixture. Ideal liquid mixture

An ideal gas mixture obeys the state equation

$$p\mathscr{V} = N\hat{R}T = (N_1 + N_2 + \ldots + N_k) \hat{R}T, \tag{104}$$

where $\hat{R} = 8.314$ J/mole-K is the universal gas constant. The partial molar volume of component j is (cp. (95))

$$\hat{\bar{v}}_j = \hat{R}T/p = \hat{v}_j, \tag{105}$$

which confirms ideality, since it is also the molar volume for the component outside the mixture at (T, p), where the partial volume is $\mathscr{V}_j$, and where it satisfies the ideal gas equation ((50)):

$$p\mathscr{V}_j = N_j\hat{R}T. \tag{106}$$

The sum of (106) written for all components shows, by comparison with (104), that the total volume of the mixture is the sum of the partial volumes of the components:

$$\mathscr{V} = \Sigma\mathscr{V}_j. \tag{107}$$

According to (106), a component fills its partial volume when it is at total pressure at the prevailing temperature. On the other hand, if it fills the total volume of the mixture, its pressure equals the partial pressure p_j, which therefore satisfies the ideal gas equation

$$p_j\mathscr{V} = N_j\hat{R}T. \tag{108}$$

When (108) is summed for all components, we obtain by comparison with (104) the total pressure as the sum of all partial pressures:

$$p = \Sigma p_j. \tag{109}$$

If we take the ratio between (109) and (104) and between (106) and (104) and use definition (63), we obtain

$$\chi_j = N_j/N = \mathscr{V}_j/\mathscr{V} = p_j/p. \tag{110}$$

Since each component, as well as the total mixture, obeys the ideal gas equation, we also have the relations (cp. Section 3.8.5)

$$\hat{u}_m = \hat{u}_m(T), \quad d\hat{u}_m = \hat{c}_{v,m}\,dT, \quad \hat{c}_{v,m} = \Sigma\chi_j\hat{c}_{v,j}, \tag{111}$$

where m denotes mixture. The enthalpy of the mixture is defined by

$$\hat{h}_m = \hat{u}_m + p\hat{v}_m = \Sigma\hat{u}_j\chi_j + p\,\hat{v}, \tag{112}$$

with

$$\hat{h}_m = \hat{h}_m(T), \quad d\hat{h}_m = \hat{c}_{p,m}\,dT, \quad \hat{c}_{p,m} = \Sigma\chi_j\,\hat{c}_{p,j}. \tag{113}$$

Changes in internal energy and enthalpy are obtained by integration of differential relations (111) and (113) respectively.

To obtain the entropy of the mixture we first determine the entropy change for each component considering the mixing process at constant T,p. This leads to result (101), or

$$\hat{\bar{s}}_j(T,p,\chi_j) = \hat{s}_j(T,p) - \hat{R} \ln \chi_j, \qquad (114)$$

which is used in (94). Dividing this by the total number of moles gives

$$\hat{s}_m = \Sigma \chi_j \hat{s}_j - \hat{R} \Sigma \chi_j \ln \chi_j, \qquad (115)$$

where $\hat{s}_j(T,p)$ for the component as a pure substance is found by integration of (44) or (48) from a given reference state.

The chemical potential $\hat{\mu}_j$ for a component in the mixture can now be calculated. From (87), and by definition, the molar Gibbs functions of the mixture is given by

$$\hat{g}_m = \Sigma \hat{\mu}_j \chi_j = \hat{h}_m - T \hat{s}_m. \qquad (116)$$

Use of (115) again gives (102), or

$$\hat{\mu}_j(T,p,\chi_j) = \hat{g}_j(T,p) + \hat{R}T \ln \chi_j, \qquad (117)$$

where

$$\hat{g}_j(T,p) = \hat{h}_j(T) - T \hat{s}_j(T,p) \qquad (118)$$

is the Gibbs function of the pure component at (T,p).

The Gibbs function for each component $\hat{g}_j(T,p)$ is often related to the value $\hat{g}_j(T,p_0)$ at the same temperature but at the standard pressure $p_0 = 1$ atm $= 1.013$ bar. When the Gibbs relation, expressed as the Gibbs function of the pure component, i.e. $d\hat{g}_j = - \hat{s}_j \, dT + \hat{v}_j \, dp$, is integrated from p_0 to p at constant T, we obtain, with $\hat{v}_j = \hat{R}T/p$,

$$\hat{g}_j(T,p) = \hat{g}_j(T,p_0) + \hat{R}T \ln (p/p_0), \qquad (119)$$

which, inserted into (117), gives

$$\hat{\mu}_j(T,p,\chi_j) = \hat{g}_j(T,p_0) + \hat{R}T \ln (p/p_0) + \hat{R}T \ln \chi_j. \qquad (120)$$

We will use this expression for calculation of the chemical equilibrium of a reactive gas mixture.

Inspection of the above results shows that all partial molar state properties for a component in an ideal gas mixture at (T,p) is equal to the specific state property of the pure component *at the temperature of the mixture and the partial pressure of the component.* For example, $\hat{\bar{h}}_j(T,p,\chi_j) = \hat{h}_j(T,p_j) = \hat{h}_j(T)$, and, in accordance with (114), $\hat{\bar{s}}_j(T,p,\chi_j) = \hat{s}_j(T,p) - \hat{R} \ln (p_j/p) = \hat{s}_j(T,p_j)$.

Calculation of the specific state properties of components in reactive gas mixtures requires the use of absolute enthalpies and entropies as references. Definition and use of these are discussed in Section 4.5.1. During processes in non-reactive mixtures, no changes in bound chemical energy take place and arbitrary reference states can be used for each component. Internal energy and enthalpy must, of course, have the same reference state for a given component.

It can be shown that results (111)–(120) are also valid for ideal mixtures of liquids. Ideal conditions for liquids also imply incompressibility, cp. Section 3.8.6.

3.10.5   Non-ideal mixtures. Activities

In the ideal mixture, the chemical potential $\hat{\mu}_j$ is calculated from (117) or (120), i.e. solely from the state properties of the component in pure form and its mole fraction χ_j in the mixture.

For non-ideal mixtures, (117) can be generalized by introducing *an activity measure* and a reference state so that the chemical potential can be calculated by integration of the expression $d\hat{\mu}_j = \hat{R}T \, d(\ln a_j)$, which gives

$$\hat{\mu}_j(T,p,\chi_j) = \hat{\mu}_j^0(T_0,p_0) + \hat{R}T \ln (a_j/a_j^0). \tag{121}$$

Here, $\hat{\mu}_j^0$ is the chemical potential for component j at the *standard activity* a_j^0, where the mixture is ideal, or hypothetically ideal, and a_j is the activity of the component in the actual mixture. The definition of $\hat{\mu}_j^0$ is thus dependent on the choice of standard state and measure of activity. These are usually defined as in Table 3.3, with temperature and pressure at the standard state given by $(T_0,p_0) = (25\,°C, 1\ atm)$.

As shown in Table 3.3, components in a gaseous, liquid or solid phase have standard states that correspond to the state of the components in pure form, $\hat{\mu}_j^0(T_0,p_0) = \hat{g}_j(T_0,p_0)$. The last term in (121) corrects this value to the actual value with respect to temperature, pressure and concentration. For example, (120) shows the contributions from deviations in pressure and the concentration for a component in an ideal gas. The contribution from deviation in temperature at constant pressure is calculated from (56) and (58) for $\hat{h}_j$ and $\hat{s}_j$, respectively, and hence for $\hat{g}_j = \hat{h}_j - T\hat{s}_j$.

For components in aqueous solutions, the standard state is chosen as the state of a 1.0 molar solution. Evaluation of the last term in (121) is discussed below.

It should be noted that expression (121), together with (96), can be

Table 3.3. *Commonly used activity measures*

Phase	Activity measure a_j	Standard activity a_j^0
gas (g)	partial pressure p_j	$p_0 = 1$ atm $= 1.013 \times 10^5$ Pa
liquid (*l*)	mole fraction χ_j	$\chi_j^0 = 1$ (pure matter)
solution in water (aq)	concentration c_j	$c_j^0 = 1$ mol/l
solid (s)	mole fraction χ_j	$\chi_j^0 = 1$ (pure matter)

used to calculate the entropy of a component at concentrations that differ from the standard state. In the case where the enthalpy change is zero, or, as is often the case, vanishing, the change in entropy is $-\hat{R} \ln (a_j/a_j^0)$. This is illustrated in Example 5.7 and corresponds to the use of (114) for an ideal gas.

In many situations, differences in the chemical potential between two phases A and B are due to differences in activity only, so that the references cancel and we have

$$\hat{\mu}_{j,\mathrm{A}} - \hat{\mu}_{j,\mathrm{B}} = \hat{R}T \ln (a_{j,\mathrm{A}}/a_{j,\mathrm{B}}) \text{ (constant } T, p). \tag{122}$$

Most living systems are isothermal but contain phases with differences in presssure. Here,

$$\hat{\mu}_{j,\mathrm{A}} - \hat{\mu}_{j,\mathrm{B}} = \hat{\mu}_j^0(T,p_\mathrm{A}) - \hat{\mu}_j^0(T,p_\mathrm{B}) + \hat{R}T \ln (a_{j,\mathrm{A}}/a_{j,\mathrm{B}})$$
$$\text{(constant } T), \tag{123}$$

where the pressure dependence of the chemical potential at the reference state must be calculated (see below).

We next illustrate the use of absolute values for the chemical potential. For an ideal mixture with $a_j = \chi_j$ and $a_j^0 = 1$ inserted into (121), we obtain (117) provided $\hat{\mu}_j^0 = \hat{g}_j(T,p)$. If the partial pressure is used as an activity measure (ideal gas mixture), so that $a_j = p_j = p \chi_j$ and $a_j^0 = p_0$ in (121), we obtain (120) because $\hat{\mu}_j^0 = \hat{g}_j(T,p_0)$. In an analogous manner, an expression for $\hat{\mu}_j^0$ can be calculated from the ideal liquid mixture with $a_j = c_j$ and $a_j^0 = c_j^0 = 1$ mol/l by converting the mole fraction χ_j^0 to a concentration c_j. It should be remembered that the value of the chemical potential $\hat{\mu}_j^0$ at standard activity depends on the chosen activity measure.

For non-ideal mixtures, $\hat{\mu}_j^0$ in (121) can be perceived as an empirical constant. Non-ideal conditions can often be treated by expressing the activity measures in Table 3.3 by *activity coefficients*, for example f_j for mole fractions, defined by

$$a_j/a_j^0 = \chi_j f_j, \tag{124}$$

or y_j for concentrations:

$$a_j/a_j^0 = c_j\, y_j. \tag{125}$$

With ideal conditions we have that $f_j = 1.0$ and $y_j = 1.0$.

For hydrogen ions, it is common practice to use either the arbitrary reference

$$\hat{\mu}^0(H^+) = 0 \text{ for } c(H^+) = 1.0\,M, \tag{126}$$

or the pH measure which is defined by

$$pH = -\log\,[c(H^+)/(1.0\,M)], \tag{127}$$

where pH = 7.0 for pure water at 25 °C = 298.15 K. Another alternative is based on reference to the electromotive force of a standard cell and a standard solution (see, e.g., Harned and Owen, 1958). Using (121), (126) and (127), the chemical potential for H^+ can be expressed in terms of the pH of the solution:

$$\hat{\mu}(H^+) = \hat{\mu}^0(H^+) + (\hat{R}T \ln 10) \times (-pH) = 5.705 \times pH \text{ kJ/mol}. \tag{128}$$

For two (aqueous) solutions separated by a membrane that is permeable to the solvent S but impermeable to the solute j it is often convenient to introduce the property called the *osmotic pressure* $\pi_{S,j}$. It is *defined*, with reference to the given binary system, as the excess pressure that must be exerted on the solution for the chemical potential of the solvent to attain the same value as the pure solvent at the same temperature. As for other properties of state, the osmotic pressure requires thermodynamic equilibrium, here between the two phases A and B, as shown in Figure 3.13. Thermal equilibrium (to be elaborated in Chapter 6) requires that

$$T_A = T_B = T, \tag{129}$$

and chemical equilibrium for component S (for which there are no constraints regarding the transfer in the composite system A + B) that

$$\hat{\mu}_{S,A}(T,p_A) = \hat{\mu}_{S,B}(T,p_B). \tag{130}$$

Substitution of (130) into (123), use of the approximation $\hat{\mu}_S^0 \approx \hat{g}_S$ and integration of the Gibbs relation, expressed by $\hat{g}$, i.e. $d\hat{g}_S = -\hat{s}_S\, dT + \hat{v}_S\, dp$ for constant T, gives

$$\int_{p_A}^{p_B} \hat{v}_S\, dp = \hat{R}T \ln\,(a_{S,B}/a_{S,A}), \tag{131}$$

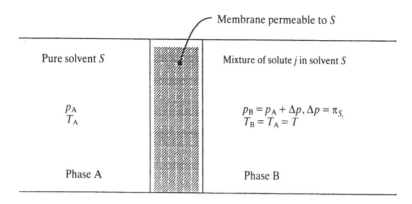

Fig 3.13 Definition of osmotic pressure $\pi_{S,j}$ at equilibrium.

or, for $v_S \approx$ constant, $a_{S,A} = a_S^0$ and $p_B - p_A = \pi_{S,j}$:

$$\pi_{S,j} = -(\hat{R}T/\hat{v}_S)\ln(a_{S,B}/a_S^0), \qquad (132)$$

from which the osmotic pressure can be calculated by use of (124) or (125). With deviations from the equilibrium $\Delta p = \pi_{S,j}$, there is transport of solvent through the membrane into the solution (or $\Delta p < \pi_{S,j}$) (osmosis) or from the solution (for $\Delta p > \pi_{S,j}$) (reversed osmosis), see, further, Section 7.2.3.

Example 3.22

The mole fraction of water in a 2.5% solution of sodium chloride (completely dissociated) is (recall Example 3.20)

$$\chi_S = 1 - 2 \times [(2.5/58.45)/(2.5/58.45 + 100/18.02)] = 0.9847. \quad (a)$$

The molar volume of water is assumed to be constant, $\hat{v}_S = 0.018 \text{ m}^3/\text{kmol}$.

If activities are approximated as mole fractions, (132) gives the osmotic pressure at 300 K as

$$\pi(H_2O,NaCl) = -(8314 \times 300/0.018) \times \ln(0.9847) = 2.14 \text{ M Pa}. \tag{b}$$

If this salt water is enclosed in a pressure container with a membrane permeable only to water, and is subjected to an excess pressure of more than about 21.1 atm, fresh water can be produced.

Example 3.23

An ideal gaseous mixture of components S and j (phase B in Figure 3.13) is separated from pure gas S by a membrane, permeable to S alone. Show that, for isothermal equilibrium, there must be an excess pressure $\Delta p = p_B - p_A$ so that the partial pressure $p_{S,B}$ is the same as p_A.

Substitution of (120) in condition (130) gives

$$\hat{g}_S(T, p_0) + \hat{R}T \ln(p_A/p_0) = \hat{g}_S(T, p_0) + \hat{R}T \ln(p_B/p_0) + \hat{R}T \ln \chi_{S,B},$$

(a)

which reduces to

$$\hat{R}T \ln[(p_B/p_A) \chi_{S,B}] = 0,$$

(b)

or, using (110), to

$$p_A = \chi_{S,B} p_B = p_{S,B}.$$

(c)

3.11   Electrochemical state properties

As shown in Section 3.4, mass possesses a content of total energy that consists of internal, kinetic and potential energy (10). When dealing with liquids that contain ions at a given electrical potential, it is often practical to combine the internal energy of a component with its electrical potential energy so as to obtain a molar *electrochemical internal energy*

$$\bar{u}_j = \hat{u}_j + z_j \mathcal{F}\mathcal{E},$$

(133)

where the last term is defined in (18). The reason for using electrochemical state properties for an ion is that its charge is inseparably associated with its mass. In a similar manner, the molar *electrochemical enthalpy* is introduced as

$$\bar{h}_j = \bar{u}_j + p\,\hat{v}_j = \hat{h}_j + z_j \mathcal{F}\mathcal{E}.$$

(134)

and a molar *electrochemical Gibbs function*, traditionally called the *electrochemical potential*:

$$\bar{\mu}_j = \bar{h}_j - T\,\hat{s}_j = \hat{\mu}_j + z_j \mathcal{F}\mathcal{E}.$$

(135)

The last-mentioned state property is of considerable practical importance for the treatment of electrochemical systems at both equilibrium (Chapter 6) and non-equilibrium (Chapter 7) in living systems. Here, spatial differences in electrochemical potentials, e.g. across cell membranes, are often the main driving forces for transport (diffusion) of ions and, indirectly, for water.

The electrochemical potential is calculated in accordance with (121) and (130) as

$$\bar{\mu}_j = \hat{\mu}_j^0 + \hat{R}T \ln (a_j/a_j^0) + z_j \mathscr{F}\mathscr{E}. \tag{136}$$

Note that $\bar{\mu}_j$ (and $\bar{u}_j$ and $\bar{h}_j$) are particular *local properties of state* and depend on the electrical potential at the actual locality, in addition to the usual intensive state properties T, p and χ_j.

Note, in addition, that the Gibbs relation (83) for a mixture can be expressed alternatively in terms of electrochemical state properties by adding the term

$$\mathrm{d}(\Sigma z_j \mathscr{F}\mathscr{E} N_j) = \Sigma z_j \mathscr{F}\mathscr{E} \, \mathrm{d}N_j \tag{137}$$

to both sides of (83), see also Guggenheim (1929) or de Groot and Mazur (1962).

3.12 Processes in living systems

Living organisms are open chemical systems. They possess energy associated with matter and they exchange energy with their surroundings more or less continuously. This energy exchange is associated with mass fluxes and with energy transport in the form of heat and work. The relative contributions of these types of energy exchange, and their direction, depend on the type of organisms (animals, plants and micro-organisms).

Within the animal kingdom, the energy transfer associated with the flow of matter into the organism is typically much larger than that out of the organism. Furthermore, the energy transfer as heat is relatively large and directed outwards, while the energy transfer as work is small and usually appears as work done on the environment. In plants, the energy added as a result of flow of matter is relatively small. The exchange of energy with the surroundings takes place in both directions and is mainly in the form of heat, with radiation for photosynthesis being the dominating part of supply. The transfer of energy in the form of work is of quantitative importance in some species, e.g. in digging, flying and swimming animals. Some of the classifications of living systems are partly based on these differences in energy flows. A typical feature is a surprising degree of stationarity of local processes and – in homeothermic animals – a remarkable degree of isothermy. Stationarity of many fundamental processes is ensured, in spite of aperiodic intake of nutrients, by accumulation/deaccumulation of matter into appropriate chemical forms.

Processes in living systems primarily involves transformation of energy in chemical bonds into heat and they take place in order to maintain a large number of more or less stationary processes. Here, chemical reactions and transport processes dominate. During unsteady states, e.g. growth, energy is used and stored in connection with processes of accumulation and structuring of matter. During the normal stationary states, however, energy storage is conserved, while the entropy of the universe increases. As a consequence, the quality of the energy (the exergy) decreases and this enables living systems to maintain spatial and temporal chemical structures despite the natural tendency of such structures to dissipate.

The structure of a living system is enormously complicated and involves a very large number of complex biochemical reactions and transport processes. From a thermodynamic point of view, however, the structure is characterized by the existence of a limited number of classes of processes. For each class of process, one or several driving forces (local differences or gradients in mechanical, thermal, chemical and electric potentials) are maintained at non-equilibrium.

The functional organization for maintaining the above-mentioned non-equilibria is remarkably constant in both animals and plants. The key to this organization is the establishment of a sustained flux of electrons from organic (typically carbohydrate) or inorganic compounds (hydrogen, hydrogensulphide or ammonia) to oxygen or to nitrate and sulphate. The flux of electrons, which takes place in a highly specialized structure, is used to maintain a flux of protons in special channels in the same structure. The energy in the proton flux is used to synthesize adenosine-triphosphate (ATP) from adenosinediphosphate (ADP) and phosphate (P). ATP is cleaved in the reverse process and the energy liberated here can be used almost universally in the organism – through a number of couplings – to maintain vital processes in its genetically determined non-equilibria.

Some remarks on reversible and irreversible processes may be helpful in this context, since the development of living systems, with an almost unbelievable degree of order in their structure, appears to contradict the common notion that natural processes tend to dissipate and eliminate differences and structures.

For example, let two bodies with different temperatures suddenly be joined in diabatic contact (state 1), while they are thermally isolated from their surrounding. Experience tells us that, after some time, the two bodies will attain a common, intermediate temperature (state 2), due to heat exchange by conduction. Clearly, the 'structure' of state 1, quantified by two different temperatures, has been dissipated (destroyed) by the natural process of heat conduction when reaching state 2, which is structureless in the sense that the two bodies now have one and the same temperature. The process is spontaneous and irreversible, as is the process of expansion of a gas into an evacuated vessel. Now, a process $1 \rightarrow 2$ is, by definition, *reversible* whenever it can be reversed, i.e. made to proceed in the opposite direction $2 \rightarrow 1$ in such a way that all initial conditions are restored at state 1, both for the actual system and for the surroundings with which it has been in contact.

All real processes are irreversible and contribute, in accordance with the second law, to an increase in the total entropy of the universe. This is true for processes in machines, in industrial plants created by human beings, and in nature, including living systems. However, the degree of irreversibility may be different for different processes.

Consider a heavy block of iron being pushed across the rough surface of a table (process $1 \rightarrow 2$). Clearly, energy is supplied in the form of work performed (frictional work). Later, this energy appears as heat in the table and the block and it is removed therefrom under isothermal conditions. The process is clearly irreversible; no one has seen the block move spontaneously back to its initial position (process $2 \rightarrow 1$) solely by the addition of heat to the block and table. If the friction is decreased (by having polished or lubricated surfaces), 'almost' no work is required and the process $1 \rightarrow 2$ approaches a reversible (frictionless) process.

The concept of reversibility is artificial. It is, however, useful in that it represents an ideal reference that allows the formulation of a rigorous thermodynamic theory with well-defined state properties. Also, it can be used as a reference for evaluation of the quality or efficiency of real processes.

Examples of sources of irreversibility include: mechanical friction between solids in sliding contact and within flowing viscous fluids; transfer (diffusion) of heat, matter and electrical charge due to finite differences in temperature, chemical potential (concentration) and electrical potential, respectively; hysteresis due to magnetization and polarization; and spontaneous chemical reactions.

As mentioned above, the ATP–ADP reaction in living systems also proceeds in the reverse direction. This, however, cannot occur without some costs (loss of exergy) and both processes taken together are irreversible. Chemical equilibria are often said to be reversible. This statement implies small displacements towards products or reactants resulting from changes in the state of the reactive mixture, e.g. by changes in pressure, temperature or chemical potentials. However, it is only in the limit of infinitesimal changes in the state that true reversibility is realized, and, in this case, the rate of the process becomes infinitesimally small. This is also true for transfer of heat, matter and charge with respect to the limit of infinitesimal driving forces and resulting fluxes.

The order that is everywhere present in the physical and biological world is due to the simultaneous occurrence of structure-building and structure-destroying processes. The former are non-linear and usually involve complex couplings between diffusive transport and chemical reactions, and they require the supply of useful energy (exergy). The latter processes are purely dissipative and can be linear.

In the following, we analyse a number of these processes qualitatively and quantitatively using the tools of thermodynamics, in particular energy conservation, entropy increases and exergy expenditure. The bioenergetic analyses are, in general, of a summarizing character and they do not deal with biochemical details, e.g. with the mechanisms of the structure-building processes.

4

Conservation of energy (the first law)

The conservation of energy (the first law of thermodynamics) is, like the conservation of mass, a general law that is independent of the particular behaviour (the constitution) of the medium considered. It is a natural law, found by observation, induction and deduction, and is valid for the macroscopic, non-relativistic systems of interest in the present context.

Starting from the conservation of energy for a closed system (control mass), we derive the more general form for an open system (control volume). The subsequent general considerations include (i) energy transformations, from one type of energy to another, in various processes (e.g. mechanical, thermal, electrical, chemical); (ii) features of energy exchange and transformations in living systems; and (iii) efficiencies of energy transformations. Thereafter, we deal with thermomechanical energy transformations for closed and open systems, including a section on processes of fluid flow, with chemical–thermal–mechanical energy transformations, and with electrochemical processes. Finally, we analyse a simplified (but in principle complete) model of a living system, including its internal structure-maintaining processes.

4.1 Energy conservation for control mass

Consider the control mass in Figure 4.1, defined as a closed system with constant mass and having total energy E, being the sum of internal, kinetic and potential energies:

$$E = U + KE + PE = Me = M(u + \tfrac{1}{2}V^2 + \varphi). \tag{3.10}$$

Then, the conservation of energy for an unsteady process is obtained from (2.5), after division by $\mathrm{d}t$ and carrying out the usual limit, as

$$\mathrm{d}[M(u + \tfrac{1}{2}V^2 + \varphi)]/\mathrm{d}t = \dot{Q} + \dot{W} \text{ (CM)}. \tag{1}$$

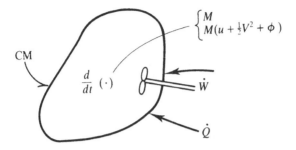

Fig 4.1 General laws for a control mass. Conservation of mass (3.7), and the first law (1).

The change over time of the total energy of the control mass is thus the sum of the instantaneous values of added powers of heat and work, where

$$\dot{Q} = \lim_{dt \to 0} \delta Q/dt \qquad\qquad \dot{W} = \lim_{dt \to 0} \delta W/dt.$$

Form (1) is the balance equation (3.2) for a control mass with $B = E$, $\dot{\Phi} = \dot{Q} + \dot{W}$ and $\dot{\Gamma} = 0$.

> Some texts on thermodynamics use the historically early sign convention in which energy as work is considered to be positive when leaving the system, hence they write $\dot{Q} - \dot{W}$ on the right-hand side of (1).

The total energy per unit of mass is the sum of internal energy u, kinetic energy $\tfrac{1}{2}V^2$ and potential energy φ. In the case of a gravitational field, (3.14) defines $\varphi = gz$, where g is the local acceleration of gravity and z the height coordinate measured upwards along the vertical z-axis. In the case of an electrical potential energy, (3.18) defines $\varphi = \hat{\varphi}_i/\hat{M}_i = z_i \mathscr{F}\mathscr{E}/\hat{M}_i$, where z_i is the ionic valence of the charged mass, $\hat{M}_i$ the molar mass, $\mathscr{F}$ the Faraday constant and $\mathscr{E}$ the electrical potential.

The heat power $\dot{Q}$ includes conduction, convection and radiation of heat to the control mass. The work power $\dot{W}$ includes mechanical work (change of volume, displacement of force, friction, etc.) and chemical work (coupling between a reaction within the system and another reaction outside the system, see Section 3.5.4). The work power associated with conservative fields of force (mass moved in a gravitational field and charge moved in an electric field) is accounted for on the left-hand side of (1) in the term for the change of potential energy, see Section 3.4.2.

It should be noted that the source term $\dot{\Gamma}$ is zero for the conservation of energy. This is also true in the case where chemical reactions (molecular or nuclear) proceed in a control mass with resulting thermal–chemical energy transformations. These transformations are described by the term $d(Mu)/dt$, where Mu is now equal to the summation $\Sigma M_i u_i$ of the internal energy for all components whose mass M_i changes during the process. Reactive mixtures of this kind are treated in Section 4.5. Also, conservation of energy is fulfilled whether or not processes in the control mass proceed through equilibrium states. However, evaluation of the internal energy u from macroscopic theory, and thereby use of (1), is possible only when the unsteady processes proceed through equilibrium states.

For a *process 1→2*, integration of (1) over time and use of $M_1 = M_2 = M$ from mass conservation, (3.8) gives

$$M[(u_2 - u_1) + \tfrac{1}{2}(V_2^2 - V_1^2) + (\varphi_2 - \varphi_1)] = {}_1Q_2 + {}_1W_2 \text{ (CM)}, \quad (2)$$

where the total energy added as heat and work is given by

$$ {}_1Q_2 = \int_1^2 Q \, dt = \int_1^2 \delta Q; \quad {}_1W_2 = \int_1^2 W \, dt = \int_1^2 \delta W. \quad (3)$$

The result (2) is valid for an arbitrary process whether or not it proceeds through equilibrium states.

For a *cyclic process*, which by definition ends at the initial state, (2) reduces to

$$0 = \oint \delta Q + \oint \delta W \text{ (CM)}. \quad (4)$$

In this section, we have used the specific energy per unit of mass, $e = E/M$. Alternatively, the molar specific energy, $\hat{e} = \hat{M}e$, can be used so that $E = N\hat{e}$, where $N = M/\hat{M}$ is the number of moles. This is shown in the next section.

4.2   Energy conservation for control volume

The first law for a control volume, defined as an open system with mass flows crossing the control surface, is obtained from (3.6) with $B = E = Me$, $b = e = u + \tfrac{1}{2}V^2 + \varphi$, $\dot{\Phi}_{CV} = \dot{Q} + \dot{W} + \dot{W}_{fl}$ and $\dot{\Gamma}_{CV} = 0$:

$$d(Me)/dt + \Sigma \, (\dot{m}e)_{\text{out}} - \Sigma \, (\dot{m}e)_{\text{in}} = \dot{Q} + \dot{W} + \dot{W}_{fl}. \quad (5)$$

This corresponds to a transformation of the left-hand side of (1) by using (3.5), and the addition of the contribution $\dot{W}_{fl}$ for the flow work that is associated with the pressure forces prevailing where mass flows cross the boundary of the control volume.

The contribution to $\dot{W}_{fl}$ from outward flow during the time interval Δt of a volume $\Delta \mathscr{V} = A\Delta l$ over the area A, on which the pressure p acts, can be calculated as the force $-pA$ times the displacement Δl in the direction of the force:

$$\dot{W}_{fl,out} = \lim_{\Delta t \to 0} [(-pA) \Delta l/\Delta t]_{out} = -(pv\dot{m})_{out},$$

where the specific volume $v = \Delta \mathscr{V}/\Delta M = A\Delta l/\Delta M$ is introduced so that the mass flow, $\lim(\Delta M/\Delta t) = \dot{m}$, appears as a factor. When the flow is directed inwards, a corresponding contribution with opposite sign is obtained since the pressure force is pA in the direction of the flow. Summation of all contributions from inward and outward flows gives

$$\dot{W}_{fl} = -\Sigma \, (pv\dot{m})_{out} + \Sigma \, (pv\dot{m})_{in}. \tag{6}$$

This result is substituted into (5) and is combined with the flow terms on the left-hand side of the equation. Using $e = u + \frac{1}{2}V^2 + \varphi$ here and introducing the definition of enthalpy $h \equiv u + pv$, the first law can now be written for a control volume as (Figure 4.2, see also Figure 1.3)

$$d[M(u + \tfrac{1}{2}V^2 + \varphi)]/dt + \Sigma \, \dot{m}(h + \tfrac{1}{2}V^2 + \varphi)_{out}$$
$$- \Sigma \, \dot{m}(h + \tfrac{1}{2}V^2 + \varphi)_{in} = \dot{Q} + \dot{W} \text{ (CV)}. \tag{7}$$

The increase over time of the total energy of the control volume plus the net outflow of enthalpy and kinetic and potential energy is equal to the added power in the form of heat and work.

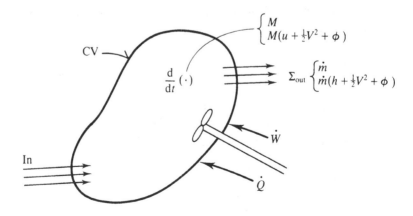

Fig 4.2 General laws for a control volume. Mass conservation (3.9) and the first law (7).

The heat power $\dot{Q}$ includes conduction, convection and radiation of heat to the control volume. The work power $\dot{W}$ includes, as for the control mass, mechanical work (volume changes against pressure forces, displacement of external force, etc.), chemical work (to be denoted $\dot{W}_{Ch}$, see Section 4.5.5), and electric and magnetic work from polarization and magnetization. However, the electric power arising from a current passing through an electrical resistor inside the control volume (Joule heating) is accounted for on the left-hand side of (7). Here, such contributions will appear as inflow and outflow of mass of electrically charged components having potential energy φ related to the electrical potentials at the points at which they cross the boundary of the control volume (cp. (3.17)).

For a *steady flow process*, the first term in (7) is zero. Furthermore, if there is only one inflow and one outflow, (3.9) shows that $\dot{m}_{out} = \dot{m}_{in} = \dot{m}$, so that (7) reduces to

$$\dot{m}[(h_{out} - h_{in}) + \tfrac{1}{2}(V_{out}^2 - V_{in}^2) + (\varphi_{out} - \varphi_{in})] = \dot{Q} + \dot{W} \text{ (CV).} \quad (8)$$

Many problems of energy analysis of living systems are of this kind and the introduction of enthalpy as the sum of the internal energy and the flow contribution is now seen to be convenient. Most tables and diagrams give values for the enthalpy h but not for the internal energy u, which must be calculated from $u = h - pv$.

It should be noted that the general form ((7)) for a control volume reduces to the special form ((1)) for a control mass, when no mass flows cross the control surface. In subsequent applications of the first law, we therefore use (7) initially and reduce it depending on the specific conditions of the problem considered.

For mixtures and for chemical reactions, it is convenient to use molar- instead of mass-based specific state properties. Introducing mole numbers $N = M/\hat{M}$ and mole flows $\dot{n} = \dot{m}/\hat{M}$, (7) takes the form

$$d[N(\hat{u} + \tfrac{1}{2}V^2/\hat{M} + \bar{\varphi})]/dt + \Sigma_{out}\,\dot{n}(\hat{h} + \tfrac{1}{2}V^2/\hat{M} + \hat{\varphi})$$
$$= \dot{Q} + \dot{W} \text{ (CV),} \quad (9)$$

where the single summation 'out' is to be understood as 'net out' (= out − in) so that energy flows into the control volume are counted as negative contributions. In the case of mixtures, the state properties in (9) are as described in Section 3.9.2, e.g. $N\hat{u} = U$, as given by (3.80). These relations are treated in more detail in Section 4.5

4.3   General considerations of energy conservation

Before illustrating the use of the first law in relation to specific problems, we make some general assumptions about the form of energy, its exchange and transformation in living systems, and associated efficiencies with respect to mechanical work and growth.

4.3.1   Energy transformations

The first law for a given control volume, expressed as (7) or (9), is a mathematical bookkeeping over contributions to changes in the energy of the system ensuring that the total energy is conserved at all times: any model for the contributions must be in accordance with the first law. Furthermore, since the individual contributions often represent different forms of energy and their sum is conserved, the result of a given process will normally imply transformations of energy. Identification and interpretation of the energy transformations is therefore an important supplement to the analysis of processes in living systems.

We have introduced the natural classification of energy forms into internal, kinetic and potential energy (associated with mass) and work and heat (associated with energy in transit). This classification is described in detail in Sections 3.4–3.6 (cp. Figure 3.11) and has been used in formulating the first law. However, it is useful to consider, furthermore, the internal energy as consisting of thermomechanical and chemical energy. This is actually suggested by the Gibbs relation for mixtures ((3.83)), where changes in internal energy can arise through separate changes in entropy (reversible heat), in volume (reversible work) and mobile number (chemical reaction). The last-mentioned process can be said to liberate or absorb (internal) bound chemical enereegy.

If, for example, a living system is considered as a whole, defined by the control surface in Figure 1.6, and processes are steady, we conclude that the difference in internal chemical energy between mass flows to and from the system is transformed into heat and work. If the processes are unsteady because chemical energy is stored or removed from internal depots, we also have chemical–chemical energy transformations. The choice of control surface determines which energy transformations will enter into the first law analysis. For example, if we choose to consider reaction (4) in Figure 1.6 as the system, we conclude that the chemical energy from the ATP→ADP reaction is transformed into internal work through setting up and relaxing the tensions in muscles.

Steady heat conduction through a wall, which defines the control mass (e.g. the blubber in Figure 3.9), is not associated with transformation of energy. The energy added to the system is equal to that removed, both as heat. If, however, the process were unsteady, there would be some energy conversion between heat and stored internal (thermal) energy as a result of a change in the temperature of the wall.

The photosynthetic process in plants involves absorption of electro-magnetic energy from the visible part of the spectrum. This energy is included here as added heat and it is transformed, by interaction with the chlorophyll molecules, into internal chemical energy, bound in carbohydrates. As in the photoelectric effect, where the absorbed radiation (accounted for as heat) is transformed into electrical potential energy (accumulation of charge over a voltage difference), there is selective absorption of photons of specific wave lengths. In contrast, absorbed thermal radiation (also accounted for as added heat) is transformed into internal (thermal) energy, which under unsteady conditions is observed as a change in temperature. Thermal radiation usually involves the exchange of energy over the complete electromagnetic spectrum.

The above assumptions are exemplified in subsequent sections, where the conservation of energy is used to analyse specific problems.

4.3.2   Energy expenditure in living systems

In the foregoing section, we introduced the term 'energy trans-formation' in order to stress the fact that many processes in living systems involve the transformation of energy from one form to another form. In addition, it is often of considerable interest to determine the energy 'costs' of maintaining the processes in parts of or in a whole organism. These costs, often called *energy expenditure*, can be defined precisely and calculated from the energy balance, whenever the conditions for the processes have been simplified.

We define *normal conditions* as those in which the organism is in a steady state or consumes (but does not accumulate) energy bound in matter in depots and gives up energy to the surroundings in the form of heat and work. In plants and some bacteria, we include the radiation energy $\dot{Q}_u$ used in photosynthesis, here identified by a separate term.

For these normal conditions, shown for the control volume in Figure 4.3, inspection suggests that terms in the energy balance (9) be collected

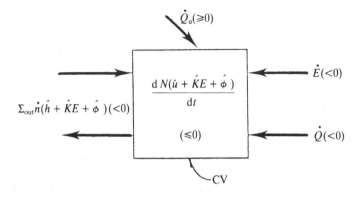

Fig 4.3 First law for living system (CV) under normal conditions.

in two groups. Using the normal sign convention, we can therefore express the energy expenditure $\dot{E}$ in two alternative forms:

$$\dot{E} = - \, \mathrm{d}/\mathrm{d}t[N(\hat{u} + \tfrac{1}{2}V^2/\hat{M} + \hat{\varphi})] \\ + [-\Sigma_{\mathrm{out}}\dot{n}(\hat{h} + \tfrac{1}{2}V^2/\hat{M} + \hat{\varphi})] + \dot{Q}_{\mathrm{u}}, \qquad (10)$$

$$\dot{E} = (-\dot{Q}) + (-\dot{W}). \qquad (11)$$

Here, the terms on the right-hand side of (10) represent consumption from depot, consumption from net intake and consumption from radiation respectively. The terms $N\bar{u}$ and $\dot{n}\hat{h}$ represent summations over all components in the case of mixtures. The terms on the right-hand side of (11) represent heat lost to the surroundings and work expended on the surroundings. In accordance with the stated conditions, all terms in (10) and (11) are positive, and the energy expenditure $\dot{E}$ can be calculated or observed from the contributions in (10) or (11) (see Section 4.7). Changes in kinetic and potential energy are usually zeero or vanishing and can be ignored. When the process is steady, the first term on the right-hand side of (10) is zero. When the external work $(-\dot{W})$ is increased, consumption from depot or from intake must increase and the energy expenditure increases.

Among conditions *not* defined as *normal*, consider the case where there is an accumulation of energy associated with matter and a simultaneous increase of body mass, i.e. $\mathrm{d}M/\mathrm{d}t > 0$ in (3.9), for example due to growth or production of a fetus, without consumption from depots. For this type of unsteady state, the first term in (9) is positive and as such contributes to the energy expenditure, which takes the forms

$$\dot{E} = [-\Sigma_{\text{out}}\, \dot{n}\, (\hat{h} + \tfrac{1}{2}V^2/M + \varphi)] + \dot{Q}_{\text{u}}, \tag{12}$$

$$\dot{E} = \text{d}[N(\hat{u} + \tfrac{1}{2}V^2/M + \varphi)]\, \text{d}t + (-\dot{Q}) + (-\dot{W}). \tag{13}$$

The same expression is valid for an animal which increases its temperature, i.e. its internal energy, after hibernating.

In each new case, expressions for the energy expenditure are derived by an appropriate grouping or partition of the terms in (9) as contributions to consumption and remaining contributions. It may be necessary to divide terms, for example so that the first term in (9), describing the time-rate-of-change of energy, consists of contributions to both consumption from and accumulation to depots.

The energy expenditure at rest $\dot{E}_0$ (Example 4.2) equals the heat production (Examples 3.2 and 3.16), since $\dot{W} = 0$. For a normal human subject ($M = 70$ kg), $\dot{E}_0$ is about 70 W, i.e. about 1 W/kg. For homeothermic animals from mice to elephants, $\dot{E}_0$ can be described, with remarkable accuracy, as a linear function of the body weight raised to the power 0.7 (cp. Example 3.12). Homeothermic animals with body temperatures between 25 and 40 °C have an $\dot{E}_0$ about five times that of poikilothermic animals of the same body weight at 15 °C, and only a part of this difference can be explained by differences in body temperature.

4.3.3 Energy efficiencies

Processes in machines or living systems can be evaluated in terms of efficiencies η of the energy transformation taking place. Such measures of performance are based on appropriate definitions and they are dimensionless.

For processes that perform work, the energy efficiency, satisfying $\eta < 1$, is normally of the form

$$\eta = \text{yield/cost}, \tag{14}$$

$$\eta = \text{actual yield/theoretically maximal yield}. \tag{15}$$

Reciprocal forms,

$$\eta = \text{theoretically minimal cost/actual cost}, \tag{16}$$

or $\eta =$ theoretical energy expenditure/actual energy expenditure, are used for processes that consume work. When these measures are based on energy, they are called *energy (or energetic) efficiencies*. Later (Section 8.3.2), we introduce efficiencies based on exergy (the Gibbs free energy).

With energy expenditures $\dot{E}_0$ and $\dot{E}$, during rest and during performance of an external work $(-\dot{W})$, respectively, cp. Section 4.3.2, we can use (14) and define a *gross efficiency*.

$$\eta\text{-gross} = (-\dot{W})/\dot{E}, \tag{17}$$

and a *net efficiency*

$$\eta\text{-net} = (-\dot{W})/(\dot{E} - \dot{E}_0). \tag{18}$$

The last-mentioned efficiency is a measure of the extra costs, i.e. the increase in energy expenditure, due to the performance of work.

The *external work* of living organisms is usually mechanical work associated with movements (of the whole body or parts thereof) against forces from the surroundings. This work can be calculated as the integral of force times displacement, cp. Section 3.5. During horizontal motion (bird in flight, swimming fish, runner), the external forces are associated with friction; during vertical motion (staircase or mountain climbing) there is, in addition, the gravitational force (a change in potential energy). Common experience shows the importance of the type of terrain (floor, gravel, sand, deep snow) on the frictional resistance for locomotion.

In contrast with the above-mentioned performance of external work, movements due to muscular activity within the organism (respiratory movements, pumping of blood, chewing of food, propulsion during peristalsis in the gut, etc.) is denoted *internal work*, as mentioned in Section 3.5.2. In this work, we must include the work of deceleration and acceleration of parts of the body during external work. The quantity of internal work cannot, in general, be observed directly. Indirectly, however, it may be determined by analysis using a proper choice of control surface around parts of the body.

In human subjects, the external work is usually quite small compared to the heat output and cannot, in practice, be calculated with any accuracy. An exception is work performed on a *bicycle ergometer*, which is often used in *ergonometrics* (the science of measuring work). Here, interest is focussed on the magnitude of the mechanical efficiency at different rates and types of external work. During work on a bicycle ergometer, the subject drives, by pedalling in the usual way, a wheel whose rotation is being braked. The cycle is calibrated so that a given speed and a given braking torque correspond to a given mechanical power. For normal subjects, the net efficiency of bicycle work is about 0.20. This net efficiency covers a number of stages of energy conversion, notably those of metabolic chemical energy into mechanical energy, and it may also be denoted the *net chemo-mechanical efficiency*. This term is conveniently used (in Example 4.9) to cover both the chemo-mechanical efficiency of the heart muscle, for the conversion of chemical energy into internal mechanical work, and the mechanical pump efficiency, for the conversion of motion of chamber and valves into useful energy received by the blood stream.

Effective growth of living systems is of importance for agriculture, fishery and forestry. In particular, in animal and fish husbandry, simultaneous observations of growth and feed intake can be used to evaluate efficiency.

By analogy with the energy expenditure at rest, the *energy expenditure of maintenance* $\dot{E}_m$ can be defined for an animal when it maintains weight and body composition under defined normal conditions. In this situation, the accumulation in (13) is zero, and $\dot{E}_m$ equals the consumption from net intake in (12), or the power removed as heat and work in (13).

Growth can be defined as the accumulation of energy in storage $\dot{E}_a$, i.e. the first term in (13). This can be determined by chemical analysis of bodies in animals, while the energy expenditure is determined by way of consumption from net intake in (12). A net efficiency for *growth* can then be defined as

$$\eta\text{-net} = \dot{E}_a/(\dot{E} - \dot{E}_m). \tag{19}$$

In this expression, all energies are so-called *metabolizable energies*. These include energies associated with intake or depots of matter that can be utilized in metabolic processes (mainly combustion) to maintain life processes. The theoretical and experimental background for the determination of these energies will be discussed in more detail later (Sections 4.5.3 and 4.5.4). The net efficiency (19) thus expresses the ratio of energy-based growth and excess energy intake. It can be obtained as the slope of an experimentally determined curve that shows the accumulated rate-of-energy $\dot{E}_a$ as a function of the rate of excess energy expenditure $\dot{E} - \dot{E}_m$ or, in the case where the energy expenditure $\dot{E}_m$ is considered to be constant, as a function of the rate of energy expenditure $\dot{E}$.

Figure 4.4 shows results of studies on adult sheep on two different feeds: dry grass of a high quality and a mixture of hay and oats (Blaxter, 1962). The slopes of the two lines show an efficiency of about 50% for the

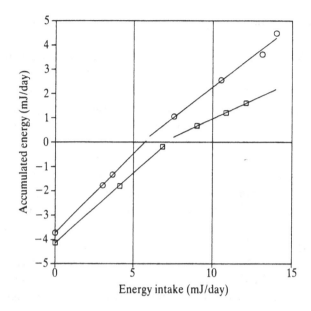

Fig 4.4 Results of energy accumulation as a function of net energy intake $\dot{E}$ from experiments with two different feeds (after Blaxter, 1962): (○) dry grass of high quality; (□) mixture of hay and oats. Note that the energy accumulation in this study is expressed as accumulation of energy of combustion and not as $\dot{E}_a$ in (13).

grass but only about 30% for the mixture of hay and oats. These efficiencies of growth are generally lower than those below the maintenance energies (58–60%). Similar studies have been performed on other animals (Kiørboe *et al.*, 1985), e.g. the common mussel *Mytilus edulis* (Riisgård and Poulsen, 1981; and Jørgensen, 1990).

Example 4.1

We wish to determine the work associated with the acceleration of a mass $M = 5\,\text{kg}$ from rest $V_1 = 0\,\text{m/s}$ to velocity $V_2 = 1\,\text{m/s}$, followed by a deceleration to rest, assuming that this process is being repeated periodically at a frequency of $f = 1\,\text{Hz}$.

For the mass as a closed system, (1) reduces to

$$d(MV^2/2)/dt = \dot{W}, \tag{a}$$

which, integrated over the period of acceleration, gives (cp.(2))

$$_1W_2 = M(V_2^2 - V_1^2)/2 = 5 \times (1 - 0)/2 = 2.5\,\text{J}. \tag{b}$$

This takes place once every second so that the external force (from muscles that move an arm holding the mass) must produce the power

$$\dot{W}_a = {}_1W_2 f = 2.5\ W. \tag{c}$$

During deceleration, (a) gives (b) with opposite sign so that the mass delivers the same energy to the surroundings (the arm). The latter absorb the energy and this is removed as heat during isothermal conditions. From a physiological point of view, the body, which delivers the force of braking, delivers the same power during periodic decelerations, $\dot{W}_d = 2.5\ W$, and the total power of the external work is

$$\dot{W} = \dot{W}_a + \dot{W}_d = 5\ W. \tag{d}$$

If we consider the mass as an external object, the body performs external work. However, the mass could also be those parts of the body that are responsible for periodic acceleration and deceleration. In this case we would consider it to be internal work.

Example 4.2

A subject performs external work with a power of 60 W and has a heat loss of 300 W during steady conditions. The energy expenditure at rest is $\dot{E}_0 = 70\ W$ and we wish to calculate the net efficiency of the mechanical work.

The total energy expenditure is (cp. (11)) $\dot{E} = 300 + 60 = 360\ W$, and (19) gives

$$\eta\text{-net} = 60/(360 - 70) = 0.207. \tag{a}$$

For comparison, (18) shows

$$\eta\text{-gross} = 60/360 = 0.167. \tag{b}$$

Example 4.3

At the beginning of the experiment described in Example 4.2, we observe that average body temperature increases steadily by 0.2 K over a period of ten minutes. The performance of external work is, however, still steady. The mass of the subject is $M = 75\ kg$, and the heat capacity of the body is assumed to be $c = 3.4\,\text{kJ/kg-K}$. Calculate the heat loss during the initial transient of ten minutes.

Because of the unsteady state, the energy expenditure (13) now contains a contribution from the change in internal energy, calculated on mass basis as

$$dU/dt = MC\,dT/dt = 75 \times 3400 \times 0.2/(10 \times 60) = 85\,\text{W}. \qquad \text{(a)}$$

It is possible that contribution (a) can be delivered as part of the heat energy that internal processes deliver for the performance of external work. If this is true, the energy expenditure, now described by (13), and the efficiencies of mechanical work will be the same as in Example 4.2.

In this case, inserting (a), $\dot{E} = 360\,\text{W}$ and $(-\dot{W}) = 60\,\text{W}$ into (13), we obtain $(-\dot{Q}) = 360 - 60 - 85 = 215\,\text{W}$. The three different states of rest, initial transient and steady state are illustrated in Figure 4.5.

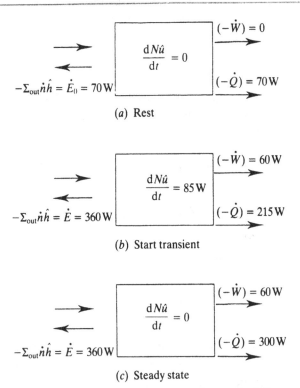

(a) Rest

(b) Start transient

(c) Steady state

Fig 4.5 Energy balances for subject (Example 4.3).

Example 4.4

For the mountaineer in Example 3.4, the term

$$\mathrm{d}(N\bar{\varphi})/\mathrm{d}t = \mathrm{d}(Mgz)/\mathrm{d}t = Mg\ \mathrm{d}z/\mathrm{d}t \qquad (a)$$

is moved to the right-hand side of (11) so that it will have a positive contribution during ascent and a negative one during descent. This corresponds to (13) and the energy expenditure becomes

$$\dot{E} = Mg\ \mathrm{d}z/\mathrm{d}t + (-\dot{Q}) + (-\dot{W}). \qquad (b)$$

Logically, the term (a) should also be added to the external work term in efficiencies such as (15) and (16), yielding, for example,

$$\eta\text{-net} = [Mg\ \mathrm{d}z/\mathrm{d}t + (-\dot{W})]/(\dot{E} - \dot{E}_0). \qquad (c)$$

The heat loss can now be calculated under the assumptions that climbing lasts for three hours, that the total energy expenditure is 6 MJ, i.e. $\dot{E} = 6 \times 10^6/3 \times 3600 = 556$ W, that the energy expenditure at rest is $\dot{E}_0 = 70$ W and that η-net $= 0.20$. Using the remaining values from Example 3.4, (a) gives

$$Mg\ \mathrm{d}z/\mathrm{d}t = 70 \times 9.81 \times 1000/(3 \times 3600) = 64\ \mathrm{W}, \qquad (d)$$

and the external work from (c):

$$(-\dot{W}) = (\dot{E} - \dot{E}_0) \times \eta\text{-net} - mg\ \mathrm{d}z/\mathrm{d}t = (556 - 70) \times 0.20 - 64$$
$$= 33\ \mathrm{W}, \qquad (e)$$

from which (b) gives the heat loss

$$(-\dot{Q}) = 556 - 64 - 33 = 459\ \mathrm{W}. \qquad (f)$$

In this example, the work needed to change the potential energy is about twice that of the frictional work. The heat loss is about 83% of the total energy expenditure. Note that external work of 97 W cannot, in general, be performed during a period of three hours without pause.

During descent, the mountaineer increases his potential energy, which may be viewed as energy received as work from the gravity field.

As for deceleration of a mass (Example 4.1), the absorption of this energy is not without cost. From personal experience, descent is strenuous.

4.4 Thermal and mechanical energy transformations

This section deals with thermal and mechanical processes in media that can be treated as a pure substance. We deal with control masses and with control volumes for which the flow of fluids (liquids and gases) is important. We also compare some actual processes to corresponding ideal processes, assumed to be reversible. To do so, it becomes necessary to draw on elements of the second law of thermodynamics, even though this law will not be formally introduced until Chapter 5.

4.4.1 Some processes in a control mass

Some of the examples related to the concept of energy in Chapter 3 will now be reconsidered and analysed here in terms of the first law and types of energy transformations. For the formulation and solution of problems, we follow the systematic procedure described in Section 2.4.

Example 4.5

A control surface around the lion, which on collision (Example 3.2) will attain zero velocity, defines it as a control mass. For the change of state $1 \rightarrow 2$ (before $\rightarrow$ after), the first law (2) gives

$$M\left[(u_2 - u_1) + \tfrac{1}{2}(V_2^2 - V_1^2)\right] = {}_1Q_2 + {}_1W_2. \tag{a}$$

While the reduction in kinetic energy is well defined and easy to calculate (Example 3.2 and 4.1), the associated energy transformation is not well defined. Inspection of (a) suggests several possibilities, e.g. (i) increase in internal energy by way of viscous dissipation at irreversible deformation of adiabatic lion; (ii) transport of heat from isothermal lion; and (iii) performance of work (force times displacement) on deformation of prey.

Example 4.6

Example 3.6 illustrates the difference between an irreversible and a reversible process for an expanding gas. In both instances, the mass of the enclosed gas is constant, cp. (3.8).

In the first case, Figure 3.7(a), the whole container is chosen as the control mass and the integrated form (2) is used on the process

$1 \rightarrow 2$, where state 2 is chosen so that all fluid flow has ended and the kinetic energy is zero. Since the change in potential energy, as well as added heat and work, is zero, (2) reduces to

$$M(u_2 - u_1) = 0. \tag{a}$$

The next step in our systematic analysis (cp. Section 2.4) is to use the Gibbs relation and equations of state. If the enclosed gas is assumed ideal, the relations in Section 3.7.5 are valid. Use of (3.56) in (a) leads to the, perhaps surprising, result that $T_2 = T_1$. The change in entropy is obtained from (3.57):

$$M(s_2 - s_1) = MR \ln (\mathcal{V}_2/\mathcal{V}_1). \tag{b}$$

Since $\mathcal{V}_2/\mathcal{V}_1 > 0$, entropy increases. This is seen as a consequence of the second law (2.6) for an adiabatic process ($\delta Q = 0$) and expresses the fact that the process is irreversible. The phenomenon is discussed in more detail in Chapter 5.

In the second case, Figure 3.7(b), only the enclosed gas is chosen as the control mass and (2) gives

$$M(u_2 - u_1) = {}_1W_2; \quad {}_1W_2 = - \int_1^2 p \, d\mathcal{V}, \tag{c}$$

where the work of expansion is assumed to be reversible and therefore calculable when the process $p = p(\mathcal{V})$ is known. Since $\mathcal{V}_2 > \mathcal{V}_1$ and $p > 0$, ${}_1W_2 < 0$, so that the internal energy must decrease. The process is an example in which internal energy is transformed into energy in the form of work leaving the system. The process is determined here by the differential form of (c):

$$M \, du = - p \, d\mathcal{V}. \tag{d}$$

Expressions for p and du, from (3.50) and (3.53), respectively, are now inserted into (d) and, after division by MRT, we obtain

$$(c_v/R) \, dT/T = - d\mathcal{V}/\mathcal{V}, \tag{e}$$

which, for constant c_v, can be integrated to

$$(c_v/R) \ln (T/T_1) = - \ln (\mathcal{V}/\mathcal{V}_1) \tag{f}$$

or

$$T/T_1 = (\mathcal{V}/\mathcal{V}_1)^{-R/c_v}. \tag{g}$$

Expressing T and T_1 by use of (3.50) gives

$$p/p_1 = (\mathcal{V}/\mathcal{V}_1)^{-(1+R/c_v)}. \tag{h}$$

We can now find $_1W_2$ by integration in (c), after inserting p from (h). However, it is easier to use the first law in (c):

$$_1W_2 = M(u_2 - u_1) = M\,c_v(T_2 - T_1), \tag{i}$$

where T_2 is given by (g):

$$T_2 = T_1(\mathcal{V}_2/\mathcal{V}_1)^{-R/c_v}. \tag{j}$$

This adiabatic process is reversible, as shown by comparison of (3.57) with (f), which shows that the process is isentropic, $s_2 - s_1 = 0$, cp. (2.6) with $\delta Q = 0$.

Example 4.7

In Example 3.9, a gas is enclosed in a container with rigid walls and heat energy $_1Q_2$ is added in a process $1 \rightarrow 2$, Figure 3.8(a). With the gas as the control mass, the first law gives

$$M(u_2 - u_1) = {_1Q_2}. \tag{a}$$

Furthermore, a process $1 \rightarrow 2'$ is described in which the same gas receives energy in the form of work $_1W_2$, Figure 3.8(b), by way of a rotating shaft equipped with a stirring device. The control surface cuts through the shaft where tensions (forces) are displaced (rotation), implying transfer of energy in the form of work. The first law now gives

$$M(u_2 - u_1) = {_1W_2}. \tag{b}$$

As noted in Example 3.8, state 1 is the same for the two cases, so whenever $_1Q_2 = {_1W_2}$, state 2 will also be the same. It is therefore impossible, on the basis of observations on state 2, to see whether the energy was added as heat or as work. In the first process, heat energy was transformed and, in the second, work energy was transformed, both into internal energy.

A closer analysis of the second process, by way of a control volume in the gas, reveals that the added work energy is first transformed into mechanical energy in fluid flows and this energy is continuously being transformed, by way of internal viscous forces, to internal energy (viscous dissipation).

4.4.2 Some processes for control volumes

Most living systems use processes in which fluid flow is important. Examples are inspiration and heating of respiratory gases, circulation of blood and filter feeding by flow through gills, see, e.g., Jørgensen (1990), Vogel (1981), Jaffrin and Shapiro (1971), Pedley (1977) and Brennen and Winet (1977).

Processes of this kind can be analysed on the basis of the first law in its general form (7) for control volumes. Often the first term in (7) is small in comparison with other terms (quasi-steady processes) or it is identically zero (steady processes), so that (7) reduces to (8). In this section, examples of the use of (8) include a heat exchanger, with humidification of air, and a pump. It is also shown when it is safe to ignore changes in kinetic energy compared to changes in enthalpy. Other problems of fluid flow will be treated in the following section.

Example 4.8

During inspiration, air flows in a quasi-steady process through the nose, pharynx and bronchi, i.e. organs which serve to absorb particles, to moisten the air with water and to exchange heat. We wish to calculate the heat added to an air flow of 6 l/min STPD when this is heated on a cold day from 0 °C to 37 °C. We first calculate this quantity by neglecting the effects of the addition of water to the air. The system is shown in Figure 4.6.

We assume the process to be steady, $\dot{W} = 0$, and the change in potential energy to be zero. Then (8) applied to the control volume in Figure 4.6 reduces to

$$\dot{m}[(h_2 - h_1) + \tfrac{1}{2}(V_2^2 - V_1^2)] = \dot{Q}. \tag{a}$$

In order to evaluate the importance of changes in kinetic energy ΔKE, we compare this with a given change in enthalpy for an ideal gas using (3.56). If the largest velocity is 14 m/s and the smallest is 0 m/s, the change in kinetic energy for air ($c_p = 1$ kJ/kg-K) will correspond to a temperature change $\Delta T = \Delta KE/c_p = (14^2 - 0)/2 \times 1000 \approx 0.1$ °C, which is negligible in this case. Thus, ΔKE in (a) can be taken as zero.

The density of air at STPD is obtained from (3.51), cp. Example 3.14, as

$$\rho_0 = 1.013 \times 10^5/(287 \times 273.15) = 1.29 \text{ kg/m}^3, \tag{b}$$

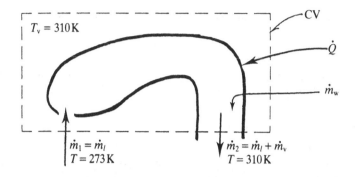

Fig 4.6 Heating of air in the nose and pharynx, with and without addition of water to reach a state of saturation (see Example 4.8).

so that (a) with the use of (3.56) gives

$$\dot{Q} = 1.29 \times (0.006/60) \times 1000 \times (37 - 0) = 4.8 \, \text{W}. \tag{c}$$

We now consider the contribution from adding water to the air. The maximum contribution is obtained if we assume that state 1 is dry air and state 2 is air saturated with water at 37 °C. The water vapour pressure under these conditions is $p_v = 6.37 \, \text{kPa}$ and this is also the partial pressure of water in the air–vapour mixture. The mass of water vapour per unit of air mass of state 2 becomes, according to (3.108) and (3.109) and at a total pressure of $p = 1.013 \, \text{atm} = 101.3 \, \text{kPa}$,

$$\dot{m}_v/\dot{m} = (M_v/M) \, [p_v/(p - p_v)] = (18.02/28.97)$$
$$[6.37/(101.3 - 6.37)] = 0.0417 \, \text{kg-vapour/kg-air}. \tag{d}$$

Conservation of mass shows that the control volume in Figure 4.6 receives the mass flow of water

$$\dot{m}_w = \dot{m} \, (\dot{m}_v/\dot{m}) = 1.29 \times (0.006/60) \times 0.0417 = 5.38 \times 10^{-6} \, \text{kg/s}, \tag{e}$$

which vaporizes so that $\dot{m}_2 = \dot{m}_1 + \dot{m}_v$, where $\dot{m}_v = \dot{m}_w$ and $\dot{m}_1 = \dot{m}_1$. The enthalpy of vaporization of water 37 °C is $\Delta h_{vw} = 2416 \, \text{kJ/kg}$ so that the heat of vaporization contributes by

$$\dot{Q}_v = \dot{m}_w \, \Delta h_{vw} = 5.38 \times 2.416 = 13.0 \, \text{W}. \tag{f}$$

The total heat power, according to (c) and (f) becomes

$4.8 + 13.0 = 17.8$ W, which is about 25% of the energy expenditure of normal subject at rest and about 10% during normal walking. A more formal analysis of the whole problem can be performed by considering the water vapour as an ideal gas and writing the conservation of mass for the control volume in Figure 4.6 as

$$\dot{m}_2 - \dot{m}_1 - \dot{m}_w = 0 \text{ or } \dot{m}_v - \dot{m}_w = 0, \tag{g}$$

and the conservation of energy as

$$\dot{m}_1 (h_2 - h_1) + \dot{m}_v h'' - \dot{m}_w h' = \dot{Q}_{tot}, \tag{h}$$

where the enthalpies of saturated vapour and liquid are h'' and h' respectively. Since both are at the same temperature $T_2 = 37\,°C$, $h'' - h' = \Delta h_{vw}$, and (g) and (h) with the use of (3.56) for air gives

$$\dot{m}_1 c_p (T_2 - T_1) + \dot{m}_w \Delta h_{vw} = \dot{Q}_{tot}, \tag{i}$$

which is the sum of (a) (for $KE = 0$) and (f).

Result (i) shows that the total heat power removed from the body in this case consists of the sum of two contributions, sensible heat (through convection to the air) and evaporative heat (through transport of water from the body, vaporizing into the air).

The first contribution, observed as an increase in the temperature of the air, is heat transfer in the sense described in Section 3.6.2. It can therefore be calculated, in principle, from temperatures, surface area, and a heat transfer coefficient for convection, according to (3.27).

The second contribution, observed as an increase in the water vapour content of the air, is a flow of enthalpy. It is difficult to calculate because it is a coupled heat and mass transfer process, relying on the supply of water and heat to produce the vapour removed by convective mass transfer in the air flow. However, given enough area and water supply, the end state becomes close to that of saturation.

Both processes of heat loss are important for the control of body temperature in living systems. Maintenance of body temperature under various conditions of energy expenditure and surroundings requires that the rates of the two processes can be controlled.

In human subjects at rest and at normal room temperature, the evaporative heat loss is about 20% of the total, but it can increase to 95% in hot and dry surroundings.

Example 4.9

The enthalpy change of a blood stream of $70 \times 70/1000 = 4.9$ l/min at an isothermal increase of pressure of 16 kPa (120 mm Hg) was calculated in Example 3.17. The increase in pressure originates in the heart pump and here we calculate the work added to the blood as useful energy.

The pump is understood as a system of channels where the movements of the walls and valves supply energy to the fluid in the form of an increase in pressure. This pressure increase ensures that the fluid can circulate through the closed circuit of blood vessels despite the associated pressure drop. The pump power $\dot{W}_p$ is supplied by a muscle (the motor), which can be analysed separately $((CV)_M$ in Figure 4.7).

Averaged over time, the process is steady so that both ΔPE and ΔKE are zero (assuming the same area and velocity in the inflow and outflow). Therefore, (8) for $(CV)_p$ in Figure 4.7 reduces to

$$\dot{m}(h_2 - h_1) = \dot{Q}_p + \dot{W}_p, \tag{a}$$

which, using (3.62) for an isothermal process, see (a) in Example 3.17, gives

$$\dot{W}_p = \dot{m}v(p_2 - p_1) - \dot{Q}_p. \tag{b}$$

So, the pump power can be calculated only when Q_p is known. The following considerations preempt the use of the second law, as described in detail in Chapter 5. There, in (5.14), we show for an *incompressible* and *isothermal* process in general, that

$$\delta Q = -\delta I, \tag{c}$$

where δI (≥ 0) denotes the total irreversibility for the differential process. Formally, (c) can be integrated for a fluid mass, or for the mass flow through the pump, to $\dot{Q}_p = -\dot{I}_p$, and (b) can be written as

$$\dot{W} = \dot{m}v(p_2 - p_1) + \dot{I}_p, \tag{d}$$

where $\dot{I}_p(\geq 0)$ is the irreversibility power (due to friction and associated internal back flow). If the pump were reversible, so that $\dot{I}_p = 0$, (d) would give

$$\dot{W}_{p,rev} = \dot{m}v(p_2 - p_1) = \dot{V}(p_2 - p_1). \tag{e}$$

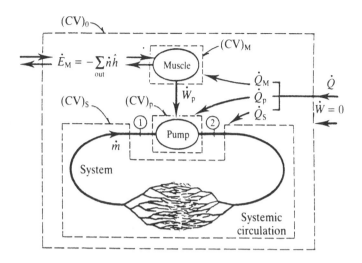

Fig 4.7 A living system with a muscle that supplies work to a pump that circulates fluid through the circulatory system.

The ratio of (e) and (d), of form (17), defines the isothermal mechanical efficiency for this pump:

$$\eta_p = \dot{W}_{p,rev}/\dot{W}_p. \tag{f}$$

To understand the meaning of mechanical efficiency fully, we note that $\dot{W}_{p,rev}$ is the useful energy received by the fluid and observed as a pressure rise, cp. (e). This energy has been received as the result of mechanical motion of the walls bounding the fluid that passes through the pump (the rotating wheel in a centrifugal pump, the undulating walls of a persistaltic motion in a gut, or the periodic contraction of the heart). The work needed to accomplish the motion of the walls is $\dot{W}_p$, which is greater than $\dot{W}_{p,rev}$ because of friction in the forced fluid motion and possible leakage in the pump. The ratio of these powers defines the mechanical efficiency, cp. (f).

In turn, the metabolic energy expenditure ($\dot{E}_M = \dot{E}_p$) needed by the muscle to effect the mechanical motion, including the friction within the muscle tissue, is greater than $\dot{W}_p$. Using (10) and (11) for $(CV)_M$ in Figure 4.7 and expecting that $\dot{Q}_M < 0$, we can calculate the total energy expenditure of the pump as

$$\dot{E}_M = -\Sigma_{out} (\dot{n}\hat{h})_M \tag{g}$$

which includes the metabolic rate at rest, a quantity that it would not be justifiable to subtract.

We can then define the metabolic or chemo-mechanical efficiency of the muscle as

$$\eta_{Ch-m} = \dot{W}_p/\dot{E}_M. \tag{h}$$

It is relatively straightforward in the present example to separate the two stages of energy conversion, as suggested in Figure 4.7. Other processes, see Example 4.15 on the ion pump, become more complex.

Combining the two efficiencies into one, representing the conversion of chemical energy into useful (reversible) energy received by the fluid, we may define a net chemo-mechanical pump efficiency:

$$\eta_{Ch-m,net} = \eta_{Ch-m} \, \eta_p = \dot{W}_{p,rev}/\dot{E}_p. \tag{i}$$

For external muscle work (ergometer bicycle experiments, cp. Example 4.2) η_{Ch-m} is about 0.2–0.25. For industrial pumps, η_p is typically 0.4–0.8. For peristaltic pumps, numerical calculations (Takabatake *et al.*, 1988) show η_p to be 0.1–0.8, depending on the amplitudes of undulations and back-pressure. Measurements of the aerobic energy expenditure of the heart in humans suggests $\eta_{Ch-m,net}$ to be about 0.2, which would be possible for, say, $\eta_{Ch-m} \approx 0.25$ and $\eta_p \approx 0.8$.

Example 4.10

A small organism has a heat loss of $(-\dot{Q}) = 1.52$ W and performs external work of 1.2 Nm/min. Calculate that part of the total energy expenditure that originates from its internal circulation that involves the pumping of 120 ml/min of fluid against a pressure drop of 25 mm Hg (3.34 kPa) with a net chemo-mechanical efficiency of 10%.

The energy expenditure of the pump is given by (g) in Example 4.9, or by

$$\dot{E} = (-\dot{Q}) + (-\dot{W}) = 1.52 + 1.2/60 = 1.54 \, \text{W}, \tag{a}$$

and we seek the part

$$\dot{E}_p/\dot{E} = (\dot{W}_{p,rev}/\eta_{Ch-m,net})/\dot{E}, \tag{b}$$

where the net efficiency is defined by (i) in Example 4.9. From expression (e) in Example 4.9, we obtain the reversible pump power as

$$\dot{W}_{p,rev} = (0.12 \times 10^{-3}/60) \times 3340 = 0.0067 \text{ W}, \tag{c}$$

and (b) gives

$$\dot{E}_p/\dot{E} = (0.0067/0.10)/1.54 = 0.0435, \tag{d}$$

which is about 4.4%.

4.4.3 Fluid flow with and without friction

In order to analyse problems of simple, steady flow in channels, we derive a simplified version of the first law here. Fluid flow in organs of living systems can be treated as incompressible, i.e. $\rho = 1/v = $ constant. This is also valid for gases, since changes in density are negligible for actual changes in pressure and temperature.

The control volume in Figure 4.8 encloses a streamtube with steady flow of an incompressible fluid without the addition of energy as work. For this control volume, (8) reduces, with $h = u + pv = u + p/\rho$, $\rho = $ constant and $\varphi = gz$, to

$$\dot{m}[(u_2 - u_1) + (p_2 - p_1)/\rho + (V_2 - V_1)/2 + g(z_2 - z_1)] = \dot{Q}, \tag{20}$$

where $\dot{m} = \rho V_1 A_1 = \rho V_2 A_2$.

To reduce this result, we use the second law, which, as shown in (5.13) for an *incompressible*, differential process in a control mass, gives

$$dU = \delta Q + \delta I, \tag{21}$$

where $\delta I(\geq 0)$ denotes the irreversibility. Following an element of mass dM from its inflow to its outflow from the streamtube in Figure 4.8, its

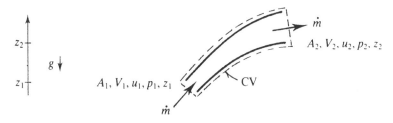

Fig 4.8 Steady flow of incompressible fluid in streamtube.

change in internal energy is $dU = (u_2 - u_1)\,dM$. Furthermore, conservation of mass ensures that the same mass dM entering the control volume also leaves it during time dt. The change over time of the internal energy for the mass dM is therefore

$$(u_2 - u_1)\lim_{dt \to 0}(dM/dt) = \dot{m}(u_2 - u_1), \tag{22}$$

and with use of the rate form of the other terms in (21) for the control volume, we have

$$\dot{m}(u_2 - u_1) = \dot{Q} + \dot{I}. \tag{23}$$

The irreversibility $\dot{I}(\geq 0)$ is associated here with viscous friction of the flow, which, according to (23), together with any added heat power is transformed into internal energy for an incompressible fluid. Equation (23) also shows that u and, in accordance with (3.60), also T, increase for an adiabatic process ($\dot{Q} = 0$), while heat must be removed ($\dot{Q} < 0$) for an isothermal process ($u_2 = u_1$).

For flow in a channel of constant diameter, fluid friction is registered as a frictional pressure drop Δp_f in the direction of the flow from 1 to 2. The irreversibility power in (23) can then be expressed as

$$\dot{I} = \dot{m}\,\Delta p_f/\rho. \tag{24}$$

Insertion of (23) and (24) into (20) gives *the mechanical energy balance for incompressible, steady flow with friction*:

$$\Delta p_f/\rho + (p_2 - p_1)/\rho + (V_2^2 - V_1^2)/2 + g(z_2 - z_1) = 0. \tag{25}$$

When the flow occurs without friction (reversibly), so that $\Delta p_f = 0$, (25) reduces to *the Bernoulli equation for incompressible, steady flow without friction*:

$$(p_2 - p_1)/\rho + (V_2^2 - V_1^2)/2 + g(z_2 - z_1) = 0. \tag{26}$$

According to (26), the sum of pressure energy p/ρ, kinetic energy $V^2/2$ and potential energy gz, all per unit of mass, is then constant throughout the fluid.

For steady flow in a channel with variable area A, conservation of mass (3.9) shows that

$$\dot{m} = \rho\,A_1\,V_1 = \rho\,A_2\,V_2 = \text{constant}, \tag{27}$$

so that areas and velocities are uniquely related.

For flow with friction in a channel with constant area, the frictional pressure drop is determined by

$$\Delta p_f = f_m \, (L/D_h) \, \rho \, V^2/2, \tag{28}$$

where L is the length of the channel and the dimensionless friction factor f_m depends on the *Reynolds number*:

$$f_m = f_m(Re); \quad Re = V \, D_h/\nu. \tag{29}$$

Here, V is the velocity, ρ the density, $\nu = \mu/\rho$ the *kinematic viscosity*, μ the ordinary *dynamic viscosity*, and D_h the *hydraulic diameter*:

$$D_h = 4 \, A/P_w, \tag{30}$$

where A is the cross-sectional area and P_w its wetted perimeter. According to definition (30), $D_h = D$ for a tube or pipe with diameter D, and $D_h = 2 \, L$ for a plane channel of width L.

For Newtonian fluids (e.g. water and air) in laminar, fully developed flow in channels ($Re < 2000$), the following theoretical results (Hagen–Poiseuille) are valid:

$$f_m = 64/Re \ \text{(pipe)}, \tag{31}$$

$$f_m = 96/Re \ \text{(plane duct)}. \tag{32}$$

For turbulent flow ($Re > 2000$) in channels with smooth walls, empirical relations, such as

$$f_m = 0.316/Re^{0.25}; \quad (10^4 < Re < 5 \times 10^5) \tag{33}$$

can be used as a good approximation irrespective of the shape of the cross-sectional area of the channel.

For non-Newtonian fluids, other relations must be used (see e.g. Yang, 1989). It is a common observation that the effective viscosity increases with decreasing velocity, as for milk products and blood, for instance. Blood is an inhomogeneous fluid consisting of suspended, deformable particles (red and white cells) in a liquid (plasma) and its flow in capillaries is complex. Thus, the effective viscosity depends on both velocity and diameter for a given vessel.

Additional frictional pressure drop occurs at the entrance to a duct, at a restriction, a branching and a change in area. These contributions are determined by empirical expressions (Fox and McDonald, 1985) and are included in Δp_f.

A typical pump–channel system (Figure 4.9) can be analysed and

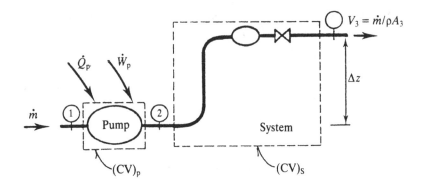

Fig 4.9 Pump and open system with friction and hydrostatic back-pressure.

described as follows. For the pump $(CV)_p$, (8) yields, using $h = u + pv$, $\Delta KE \sim 0$ and $\Delta PE \sim 0$,

$$\dot{m}[(u_2 - u_1) + v(p_2 - p_1)] = \dot{W}_p + \dot{Q}_p, \tag{34}$$

or, using (23) and the efficiency η_p from (f) and (g) in Example 4.10,

$$\dot{W}_p = \dot{W}_{p,rev}/\eta_p; \quad \dot{W}_{p,rev} = \dot{\mathcal{V}}\Delta p_p, \tag{35}$$

where $\dot{\mathcal{V}} = \dot{m}v = \dot{m}/\rho$ is the volume flow and $\Delta p_p = p_2 - p_1$ is the pressure rise. For the channel system $(CV)_s$ in Figure 4.9, (25) gives

$$\Delta p_s = \Delta p_f + \rho(V_3^2 - V_2^2)/2 + \rho g(z_3 - z_2), \tag{36}$$

where $\Delta p_s = p_2 - p_3$ is the total pressure change. Parts of, or all of, the channel system could, of course, have been placed upstream of the pump.

From (28), it can be seen that Δp_f, just as ΔKE and therefore Δp_s, depends on the volume flow. This dependency, given by (36), expresses the *system characteristic*

$$\Delta p_s = \Delta p_s(\dot{\mathcal{V}}), \tag{37}$$

which, as shown schematically in Figure 4.10, for any given system, exhibits an increasing pressure drop with increasing volume flow. In a closed system (circulation), $(CV)_s$ in Figure 4.7, we have $\Delta z = 0$ and $\Delta KE = 0$ if the cross-sectional areas are the same at inflow and outflow of the pump. When further resistance is introduced into the system, e.g. by closing a valve further, Δp_f will increase and the system characteristic will shift upwards. The same occurs if Δz is increased in an open system.

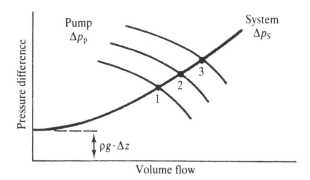

Fig 4.10 Change in pressure as function of volume flow for a given system (system characteristic) and for a pump at three activity levels (three pump characteristics). Points of intersection 1, 2 and 3 give the corresponding operating points.

A pump that works at a given rhythm (stroke frequency, stroke volume, rpm) yields a volume flow which depends on the pressure rise supplied. This dependency is expressed in the *pump characteristic*

$$\Delta p_\mathrm{p} = \Delta p_\mathrm{p}(\dot{V}). \tag{38}$$

It is determined by the type and actual design of the pump and it usually shows a decreasing $\dot{V}$ (e.g. due to increasing leakage) with increasing Δp_p. Figure 4.10 illustrates (38) for three different rhythms (activity levels) of a pump. When a pump is coupled to a system, the volume flow adjusts itself so that

$$\Delta p_\mathrm{p} = \Delta p_\mathrm{S}. \tag{39}$$

This condition corresponds to the intersection of the two characteristics and is denoted the *operating point*, cp. Figure 4.10.

The above assumptions show that the pump power can be determined, either according to (35), by observation of the volume flow and pressure increase (and knowledge of the efficiency), or by theoretical predictions based on a model of the system and pump characteristics, and the use of (39). These approaches are described for the gill pump of *Mytilus edulis* (the common mussel) in Jørgensen (1990).

Example 4.11

Several well-known phenomena follow from (26).

For a fluid at rest $(V = 0)$, the hydrostatic pressure increases with depth:

$$p_2 = p_1 + \rho g(z_1 - z_2), \tag{a}$$

which is used in liquid manometers, for instance, for determining differences in pressure. Thus, $\Delta z = z_1 - z_2 = 100\,mm\,H_2O$ corresponds to

$$\Delta p = p_2 - p_1 = 1000 \times 9.81 \times 0.1 = 981\,Pa = 0.981\,kPa, \tag{b}$$

and 120 mm Hg systolic blood pressure (Example 4.9) corresponds to

$$\Delta p = 13\,600 \times 9.81 \times 0.12 = 16\,010\,Pa = 16\,kPa. \tag{c}$$

Suppose that water (rather than air) covers the mercury in a U-tube manometer, indicating a level difference of 100 mm. Since it is the difference in hydrostatic pressure of the two columns that determines the pressure difference, we now obtain

$$\Delta p = [\rho(Hg) - \rho(H_2O)]\, g\, \Delta z$$
$$= (13\,600 - 1000) \times 9.81 \times 0.1 = 12.36\,kPa. \tag{d}$$

For flow in a horizontal channel $(\Delta z = 0)$ whose cross-sectional area decreases (nozzle, contraction), the velocity will increase according to (27) and the pressure will decrease according to (26), provided frictional forces are neglected. This principle is used in flow meters. Here a decrease in area is followed by an increase in area. In this way the pressure drop due to acceleration is followed by a pressure increase due to deceleration. The pressure recovery is complete when frictional forces are negligible.

Example 4.12

The gills in *Mytilus edulis* (the common blue mussel) consist of many parallel and plane interfilamental canals with a typical width of $l = 0.04\,mm$ and length $L = 0.2\,mm$. The flow of water has a mean velocity of about $V = 2\,mm/s$ and is driven by cilia pumps (beating lateral cilia downstream of the entrance to these channels). The water flow from all the channels is collected in a chamber, from which it leaves the animal as a jet with a velocity of about 86 mm/s. This arrangement ensures that the water, having been filtered for microorganisms, does not readily reach the intake.

Assuming the kinematic viscosity of sea water at 15 °C to be $\nu = 1.1 \times 10^{-6} \, m^2/s$, we seek the frictional pressure drop in the channels and the change in kinetic energy, i.e. the two most important contributions to the total pressure drop.

For the plane channel, (30) and (29) give

$$D_h = 4l \times 1/(2 \times 1) = 2l, \tag{a}$$

$$Re = 2l \, V/\nu = 2 \times 0.04 \times 2 \times 10^{-6}/1.1 \times 10^{-6} = 0.146, \tag{b}$$

so that the flow is laminar and dominated by viscous forces. The flow is now assumed to be fully developed and we substitute (3) into (28), which, with (a) and (b), gives

$$\Delta p_f = 12 \, \rho \, \nu \, V \, L/l^2 = 3.3 \, Pa = 0.33 \, mm \, H_2O. \tag{c}$$

The kinetic energy at the outflow corresponds to the pressure change

$$\rho \, V_3^2/2 = 1000 \times 0.085^2/2 = 3.6 \, Pa = 0.37 \, mm \, H_2O, \tag{d}$$

which is slightly larger than the pressure drop due to friction. Note that the pump must accelerate the water from the velocity of the surroundings, which is equal to zero, so that $V_2 = 0$ in (36), cp. the schematic diagram in Figure 4.7, which can be used even though the pump is placed in the channel system.

Furthermore, $\Delta z = 0$ in (36) so that (37) consists of a linear and a quadratic term. A more detailed calculation (Jørgensen et al., 1986) shows that there are additional frictional contributions so that the normal operating point for a standard mussel of length 35 mm corresponds to

$$\Delta p_S \sim 1 \, mm \, H_2O = 9.81 \, Pa; \quad \dot{\gamma} \sim 1 \, ml/s. \tag{e}$$

Using (35), we obtain the reversible pump power as

$$\dot{W}_{p,rev} = \dot{\gamma} \Delta p_p = 10^{-6} \times 9.81 \sim 0.01 \, mW. \tag{f}$$

The work done by the pump may be compared to the total energy expenditure of the mussel, as expressed by the rate of oxygen consumption. According to Hamburger et al. (1983) the 35 mm mussel, dry mass of soft tissue amounting to 0.21 g, consumes

0.169 ml/h of oxygen. If we assume that 1 ml of oxygen corresponds to 19.2 J (or 430 kJ/mol O_2, cp. Table 4.1), the aerobic metabolic rate of the mussel becomes 0.901 mW, of which the reversible pump work then constitutes 1.1%. Recalling (i) in Example 4.9 and assuming a net chemo-mechanical efficiency of 10%, the actual energy expenditure would be about 11% of the total.

Example 4.13

In human subjects, the blood circulates, starting from the single artery aorta $(D_a \sim 26\,\text{mm})$, through arteries which successively divide into smaller and smaller vessels, down to capillaries $(D_c \sim 0.007\,\text{mm})$. We seek a relation between the number of capillaries, their cross-sectional area and pressure gradient.

Assuming laminar flow of a Newtonian fluid, (28) and (31) show that

$$\Delta p_f/L \sim VD^2, \tag{a}$$

and (27) can be expressed by

$$\dot{m} = N \rho AV \sim N D^2 V = \text{constant}, \tag{b}$$

where N is the number of parallel channels which at any level of branching must carry the total volume flow.

From (a) and (b) we see, first, that the pressure gradient in a single channel $(N = 1)$ depends on the diameter as

$$\Delta p_f/L \sim 1/D^4. \tag{c}$$

The pressure drop over a given channel length will therefore increase by a factor of 16 when the diameter is halved for the same volume flow.

For N parallel channels, (a) and (b) show that

$$(\Delta p_f/L) \sim 1/ND^4. \tag{d}$$

Assuming a number of capillaries of $N_c = 5 \times 10^9$ with diameter $D_c = 0.007\,\text{mm}$, (d) gives

$$(\Delta p_f/L)_c/(\Delta p_f/L)_a = (1 \times 26^4)/(5 \times 10^9 \times 0.007^4) \approx 40\,000, \tag{e}$$

and, since $NA \sim 1/D^2$, the total cross-sectional flow area becomes

$$(NA)_c = 5 \times 10^9 \times (\pi/4) \times (7 \times 10^{-6})^2 \approx 0.2\,\text{m}^2. \tag{f}$$

Result (f) appears reasonable, while (e) can be understood only when we consider the very short length of the capillaries (about 1 mm). The total pressure drop through the circulation at rest is the same as the pressure increase of the pump, about 100–120 mm Hg. Of this, about 30% occurs in the capillaries. The blood behaves as a non-Newtonian fluid but we do not calculate the pressure drop here.

The volume of tissue perfused by capillaries in human subjects is about $30 \, l = 0.03 \, m^3$. The average distance between capillaries is therefore

$$L_{ca} \sim (\mathscr{V}_{tissue}/N)_c^{1/3} \sim (0.03/5 \times 10^9)^{1/3} = 0.182 \times 10^{-3} \, m. \qquad (g)$$

This arrangement allows the transport of matter to and from cells in the tissues by diffusion over distances of only about $(0.182 - 0.007)/2 \sim 0.09 \, mm$, i.e. typically 4–5 cell diameters.

4.5 Chemical–thermal–mechanical energy transformations

This section deals with processes in which chemical reactions occur. As mentioned in Section 3.12, such processes are the most important ones in living systems. It is through chemical reactions that the necessary form and amount of energy is delivered at the right time and place in the system for the maintenance of all the processes of life. The processes can be summarized as serving the purpose of maintaining mechanical, thermal, chemical and electrical non-equilibrium structures (the dissipative structures of life, see Section 8.6).

The basic description of mixtures is given in Section 3.9. This includes reaction equations and conservation of mass ((3.73)), which, for each of the k components in the mixture, can be written as

$$dN_i/dt + \Sigma_{out} \, \dot{n}_j = \dot{N}_j. \qquad (40)$$

Here the summation 'out' (cp. (9)) includes the molar flows 'into' the control volume as negative contributions, and $\dot{N}_j$ is the source term ((3.80)) originating from chemical reactions. In addition to (40), we need the conservation of energy ((9)), written in a simplified form here as

$$d(\Sigma N_j \, \hat{u}_j)/dt + \Sigma_{out}(\Sigma \, \dot{n}_j \, \hat{h}_j) = \dot{Q} + \dot{W}, \qquad (41)$$

where changes in kinetic and potential energy are omitted because such changes are zero or negligible for the chemical processes of interest. An exception is electrical potential energy, which is important in electro-chemical processes (see Section 4.6).

Due to chemical reactions, the mass of each component is no longer conserved. It is therefore necessary to introduce a common reference for the state properties internal energy and enthalpy, which leads us to use their absolute values. Following their introduction, we deal first with steady combustion reactions and then with unsteady reactions, which, with few assumptions, can be generalized to a simple expression for the energy expenditure of a system. On this basis and for classes of biological fuels, we show that the energy expenditure of a living system under normal conditions can be calculated from observations of uptake of oxygen and excretion of carbon dioxide and nitrogen by way of so-called 'indirect calorimetry'.

4.5.1   Absolute values of state properties

During chemical reactions, the mass of a single component is not conserved. This necessitates the use of absolute values for energy and entropy, and for the derived state properties enthalpy, Helmholtz and Gibbs functions. The procedure is to refer all state properties to one and the same standard state, by international convention fixed to the following convenient value:

$$T_0 = 25\,°C = 298.15\,K \text{ and } p_0 = 1\,atm = 101.3\,kPa. \qquad (42)$$

The enthalpy of formation, $h_f^0 = \hat{h}(T_0,p_0)$, or $\hat{h}^0$, is, by definition, equal to *zero* for *elements* of pure matter in their normal molecular form and phase, and it is equal to the enthalpy change needed for the formation of chemical compounds, i.e. the amount of energy that must be added for the formation of compounds from the constituent elements in their normal molecular form at reference state (T_0,p_0). The value of $\hat{h}_f^0$ is therefore negative for stable chemical compounds at (T_0,p_0). The term 'chemical compounds' also includes non-equilibrium states for elements with respect to their molecular form (e.g. atomic oxygen), or with respect to their crystal form or phase (e.g. diamond).

Appendix C.1 gives values of $\hat{h}^0$ for selected substances. The value of $\hat{h}^0$ for molecular oxygen (O_2) is zero, while that for carbon dioxide (CO_2) is negative, $-393\,kJ/mol$, since energy is liberated during formation: $C + O_2 \rightarrow CO_2$.

The molar enthalpy $\hat{h}(T,p)$ at another state is obtained by adding to $\hat{h}^0$

the integral of (3.47) from (T_0,p_0) to (T,p). This is given by (3.56) in the case of an ideal gas and by (3.62) in the case of an ideal liquid or solid. The molar internal energy is calculated from

$$\hat{u} = \hat{h} - p\hat{v}. \tag{43}$$

The product $p\hat{v}$ is generally very small in comparison with the enthalpy of formation for compounds in biological systems (less than 1–2% for gases and about 1000 times less for liquids and solids). We thus often use the approximation

$$\hat{u} \approx \hat{h}. \tag{44}$$

Absolute entropy at state (42), $\hat{s}^a = \hat{s}(T_0,p_0)$, is calculated by integration of the state equation from absolute zero. This follows from the third law, which states that the entropy of pure matter is zero at $T = 0$ K. The background for calorimetric and spectroscopic determination of absolute entropy is described by Prigogine and Defay (1954). Appendix C.1 gives values of s^a for selected substances.

The absolute value for molar entropy $\hat{s}(T,p)$ at another state is obtained by adding to $\hat{s}^a$ the integral of (3.48) from (T_0,p_0) to (T,p). This is given by (3.58) in the case of an ideal gas and by (3.61) for an ideal liquid or solid. Contributions from entropies of mixing are given by (3.101).

The above considerations provide a common, absolute reference state for all energetic state properties, including the molar Gibbs function $\hat{g}^a = \hat{h}^0 - T\,\hat{s}^a$. Many tables of state properties (see e.g. Schaefer and Lax, 1961; Lide, 1990) contain values both of an absolute entropy $\hat{s}^a$ and of a *standard entropy* $\hat{s}^0$ for elements and chemical compounds as pure matter at the standard state $(T_0,p_0) = (25\,°C,\ 1\ atm)$, as well as a component in aqueous (aq) solution at the standard state and concentration of 1 mol/l. Here, the standard entropy (entropy of formation) is defined as the absolute entropy of the substance minus the sum of the absolute entropies of those elements that form the substance. This implies that the standard entropy for elements is zero by definition, just as is the enthalpy of formation. It follows that the reaction entropy ((5.25)) can be expressed as $\Delta\hat{s}_r^0 = \Delta\hat{s}_r^a$ and that the reaction Gibbs function ((5.30)) is $\Delta\hat{g}_r^0 = \Delta\hat{g}_r^a$. Tables often give the standard (formation) Gibbs function $\hat{g}^0 = \hat{h}^0 - T\,\hat{s}^0$, which differs from the absolute Gibbs function $\hat{g}^0 = \hat{h}^0 - T\,\hat{s}^a$. Appendix C.1 gives values of $\hat{h}^0$, $\hat{g}^0$ and $\hat{s}^a$ for selected substances.

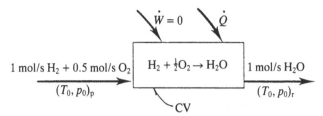

Fig 4.11 Formation of H_2O in a steady control volume process at (T_0, p_0).

Example 4.14

The significance of enthalpy of formation of the chemical compound water is shown in Figure 4.11.

1 mol/s H_2 (g) and 0.5 mol/s O_2 (g) react during isothermal and isobaric conditions at (T_0, p_0) to form 1 mol/s H_2O (g) in a steady flow process. The reaction equation fulfils the conservation of mass and, since $d/dt = 0$ and $\dot{W} = 0$, (41) reduces to

$$(\dot{n}\,\hat{h})_{H_2O} - (\dot{n}\,\hat{h})_{H_2} - (\dot{n}\,\hat{h})_{O_2} = \dot{Q}, \tag{a}$$

which using Appendix C.1 gives

$$\dot{Q}/\dot{n}(H_2O) = \hat{h}^0(H_2O) - \hat{h}^0(H_2) - \tfrac{1}{2}\hat{h}^0(O_2) = \hat{h}^0(H_2O)$$
$$= -242 \text{ kJ/mol}. \tag{b}$$

The energy supplied as heat per mole H_2O formed at (T_0, p_0) equals the enthalpy of formation. This is negative since the process is exothermic. If the product is H_2O (*l*), Appendix C.1 gives the value -286 kJ/mol, which deviates from the value for H_2O (g) by the enthalpy of evaporation at 25 °C: $\Delta\hat{h}_{fg} = 44.0$ kJ/mol. The molar internal energy for the two states is calculated from (43), where the molar volume in the *pv*-contribution is determined for the incompressible liquid (*l*) and the vapour (considered as an ideal gas, $\hat{v}(g) = 8.314 \times 298/1.013 \times 10^5 = 0.0245 \text{ m}^3/\text{mol}$) respectively. This gives

$$(H_2O)\ (g) : p\hat{v} = 101.3 \times 0.0245 = 2.5 \text{ kJ/mol}, \tag{c}$$

$$(H_2O)\ (l) : p\hat{v} = 101.3 \times (0.001 \times 0.018) = 0.002 \text{ kJ/mol}, \tag{d}$$

so that approximation (44) gives an error for the gas phase of only 1%. The error becomes smaller for larger molecules having a numerically larger enthalpy (see Appendix C.1) and it is always less for liquid and solid phases.

4.5.2   Steady combustion reaction

An important class of chemical reactions in living systems is the combustion of organic matter, mainly carbohydrate, fat and protein, with oxygen. The reactions proceed through a large number of intermediate reactions, usually catalysed by enzymes and involving several steps that are coupled to other reactions. Also, the reactions normally proceed under isothermal and isobaric conditions, i.e. $37\,°C \sim 310\,K$ and $1\,atm$. Since the temperature correction for $\hat{h}$ for deviations from the reference state (42) is small in comparison with $\hat{h}$, the value of $\hat{h}^0$ is often used (see Example 4.15).

The combustion reaction in the open system shown in Figure 4.12 can be described by *the net reaction*

$$S + O_2 \rightarrow P_1 + P_2 + \ldots, \tag{45}$$

where S denotes substrate (fuel), O_2 oxygen, which is added to the system by flow or diffusion, and P_1, P_2,... the end products which are eliminated from the system by other non-reactive processes. In addition, as also shown in Figure 4.12, the process may involve intermediary products X_1, X_2,..., and reactions $Y_1 \rightleftarrows Y_2$ which are coupled to the intermediary steps.

It is generally assumed that the processes in the reaction system of Figure 4.12 can be considered as *steady* over time periods of practical interest. This fact is explained by the operation of internal controls of the rates of several of the reactions. Also, the change over time of the amounts and energies of the intermediary components in the system are often quite small in comparison to the net contributions of the reaction (45).

As a fundamental example of (45), we consider the complete combustion of glucose (G) with oxygen, characterized by the stoichiometric reaction equation (3.75):

$$C_6H_{12}O_6 \ (aq) + 6\ O_2 \ (g) \rightarrow 6\ CO_2 \ (g) + 6\ H_2O \ (l). \tag{46}$$

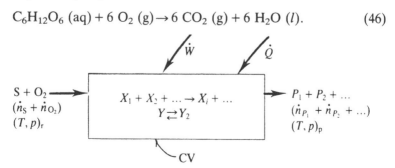

Fig 4.12 Generalized combustion reaction in an open system (CV).

For a steady process in the control volume of Figure 4.12, conservation of energy (41) gives

$$-(\dot{n}\hat{h})_G - (\dot{n}\hat{h})_{O_2} + (\dot{n}\hat{h})_{CO_2} + (\dot{n}\hat{h})_{H_2O} = \dot{Q} + \dot{W}. \qquad (47)$$

Stoichiometric coefficients v_j are related to mole fluxes by

$$\dot{n}_j/\dot{n}_G = v_j, \qquad (48)$$

and are positive for products and negative for reactants, according to (3.77). Using (48), we may write (47) as

$$\dot{n}_G \, \Delta\hat{h}_r = \dot{Q} + \dot{W}, \qquad (49)$$

or

$$\dot{n}_G \, (-\Delta\hat{h}_r) = (-\dot{Q}) + (-\dot{W}). \qquad (50)$$

Here, the molar *reaction enthalpy* (the negative of the *molar combustion enthalpy*) is defined by

$$\Delta\hat{h}_r = \Sigma v_j \, \hat{h}_j, \qquad (51)$$

where $v_j = -1$ for $j = $ fuel. The commonly used term *heating value*, defined as the heat energy given off to the surroundings under specified conditions (e.g. $\dot{W} = 0$), is identical to the enthalpy of combustion and to the negative of the enthalpy of reaction. At reference state (42), $\Delta\hat{h}_r^0 = \Delta\hat{h}_r(T_0, p_0)$ is denoted the standard reaction enthalpy. Values for selected reactions are given in Appendix C.2. For the case of (46), (51) takes the explicit form

$$\Delta\hat{h}_r = 6 \, \hat{h}_{CO_2} + 6 \, \hat{h}_{H_2O} - \hat{h}_G - \hat{h}_{O_2}. \qquad (52)$$

$\Delta\hat{h}_r$ is negative for exothermic reactions, where chemical energy is liberated and, for steady processes, must be removed. Here, form (50) is useful and its right-hand side is also the energy expenditure of the system under normal conditions, cp. (11).

It should be noted that the assumption of stoichiometric mole fluxes does not introduce restrictions for isothermal processes. If, for example, a larger flux of oxygen were to enter the sytem, the excess would leave the system with unchanged enthalpy and the energy balance would be unaffected.

For steady conditions, the mole balance (40) for component j reduces to

$$\Sigma_{out} \, \dot{n}_j = \dot{N}_j, \qquad (53)$$

or, for a simple reaction, after introduction of the *molar reaction rate* ξ from (3.89), to

$$\Sigma_{\text{out}}\ \dot{n}_j = \nu_j \dot{\xi}. \tag{54}$$

For (46), this implies four equations:

$$\left.\begin{array}{ll} -\dot{n}_G = -\dot{\xi}, & -\dot{n}_{O_2} = -6\,\dot{\xi} \\[4pt] \dot{n}_{CO_2} = 6\,\dot{\xi}, & \dot{n}_{H_2O} = 6\dot{\xi} \end{array}\right\}, \tag{55}$$

where the reaction rate $\dot{\xi}$ has unit moles of glucose per unit time. Inserting $\dot{n}_G$ from (55) into (49), we obtain

$$\dot{\xi}\, \Delta \hat{h}_r = \dot{Q} + \dot{W}, \tag{56}$$

which, as shown in the following section, is also valid for unsteady processes.

Example 4.15

We wish to calculate the reaction enthalpy for isothermal and isobaric combustion of glucose at 37 °C and 1 atm. Furthermore, we seek the energy expenditure for a steady process involving combustion of 389 g/day.

In order to calculate these quantities, we need to specify the details of reaction (46). If the control volume is assumed to be the whole organism (e.g. a human subject), the supply of oxygen and the removal of carbon dioxide will take place as gases (g), while the glucose could be added as a solid (s) or in aqueous solution (aq). If the control volume is a piece of tissue, the reaction will be considered to take place in an aqueous solution (aq) and the concentrations must be specified. In this case, we consider an aqueous solution of 0.01 mol/l glucose, having partial pressures of carbon dioxide and oxygen of 0.07 atm and 0.21 atm respectively. The latter information is without consequence for enthalpies, since we assume ideal conditions, but it will be used in the calculation of entropy and the Gibbs function (Example 5.7).

We now deal with the first condition, that of oxygen and carbon dioxide in gas phase (g) and glucose in an aqueous solution (aq). Introducing the associated enthalpies of formation from Appendix C.1 into (52), we obtain the enthalpy of the reaction at the standard state $(T_0, p_0) = (25\,°C, 1\ \text{atm})$:

$$\begin{aligned} \Delta \hat{h}_r^0 &= 6 \times (-393) + 6 \times (-286) - (-1264) - 6 \times (0) \\ &= -2810\ \text{kJ/mol}. \end{aligned} \tag{a}$$

Inspection of Appendix C.1 shows that the difference in enthalpy between solid and dissolved glucose is -11 kJ/mol. Furthermore, the differences in enthalpies of formation for gas and dissolved matter is 20 kJ/mol for carbon dioxide and 10 kJ/mol for oxygen. When (a) is corrected for these differences (or using enthalpies for all components in aqueous solutions in (52)) we obtain

$$\Delta \hat{h}_r^0 = -2870 \text{ kJ/mol (all components in aqueous solution)}. \quad \text{(b)}$$

Comparion of (a) and (b) shows that the total contribution to the change in reaction enthalpy by transfer of gaseous components to aqueous solutions is relatively small in this case, 60/2810, i.e. about 2%.

We now calculate the reaction enthalpy at state $(T,p_0) = (37\,^\circ\text{C}, 1 \text{ atm})$ by correction for the temperature dependence of the individual enthalpies. This is calculated by the use of (3.56) and (3.62) and by assuming ideality:

$$\hat{h} = \hat{h}^0 + \hat{c}_p(T - T_0), \quad \text{(c)}$$

which, for an arbitrary reaction, leads to

$$\Delta \hat{h}_r = \Delta \hat{h}_r^0 + (T - T_0) \, \Sigma \nu_j \, \hat{c}_{pj}. \quad \text{(d)}$$

Assuming the heat capacities for oxygen and carbon dioxide to be the same in gas phase and aqueous phase, we have the heat capacities from Appendix C.1 as $\hat{c}_p = 0.037, 0.075, 0.219$ and 0.029 kJ/mol-K for carbon dioxide, water, glucose and oxygen, respectively, and obtain

$$\Delta \hat{h}_r = -2870 + (37 - 25) \times 0.279 = -2870 + 3.3$$
$$= -2867 \text{ kJ/mol}.$$
$$\text{(e)}$$

This may be compared to the value in Appendix C.2, or on a mass basis,

$$\Delta h_r = \Delta \hat{h}_r / \hat{M}_G = -2867/180 = -16 \text{ kJ/g}. \quad \text{(f)}$$

The correction for this small deviation from T_0, typical of biological systems, is less than 0.2% and is often neglected. The energy expenditure at a glucose consumption of 389 g/d, or

$$\dot{n}_G = \xi = (389/180)/86\,400 = 0.025 \text{ mmol/s} \quad \text{(g)}$$

is obtained from (50) and (11), with (e) and (g), as

$$\dot{E} = \dot{n}_g\,(-\Delta\hat{h}_r) = 0.025 \times 2867 = 71.7\,\text{W}. \tag{h}$$

Note that the assumption of stationarity allows calculation of $\dot{E}$ from observation of only one of the four molar fluxes.

4.5.3 Unsteady reactions. Generalization

The results of the preceding section can be generalized for the isobaric, isothermal case when there is an accumulation (deaccumulation) over time of the components of the reaction in the control volume, Figure 4.13. However, it is still necessary to assume steady state with respect to the intermediary products X_i and their associated coupled reactants and products Y_i, which remain in the control volume, cp. Figure 4.12.

Let us use approximation (44) and assume constant (T, p), i.e. $\hat{u}_j \approx \hat{h}_j =$ constant. This simplifies the first term in (41). Let us also introduce the reaction rate into (40), assuming a single reaction. This gives

$$\Sigma\hat{h}_j\,dN_j/dt + \Sigma_{\text{out}}\,\Sigma\dot{n}_j\hat{h}_j = \dot{Q} + \dot{W}, \tag{57}$$

$$dN_j/dt + \Sigma_{\text{out}}\,\Sigma\,\dot{n}_j = \nu_j\,\xi. \tag{58}$$

In (57), and in the following, $\hat{h}_j$ is used as for ideal mixtures. Changes in the composition of biological mixtures are often small and since changes in $\hat{h}_j$ with changes in the composition of such mixtures are often small, deviations from non-ideal conditions can be accounted for by using constant average values for $\hat{h}_j$.

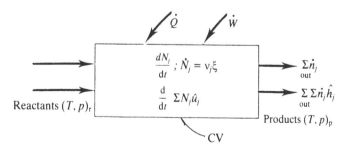

Fig 4.13 Unsteady chemical reaction. Mole balance for component j and the first law for the complete mixture.

Now, it will be recalled that Σ_{out} refers to net outflows (i.e. the sum of all outflows (positive) and all inflows (negative), of which there may be several), and Σ refers to components (i.e. summation over $j = 1,2,. . .,k$). Multiplication of (58) by $\hat{h}_j$ and summation gives

$$\Sigma\hat{h}_j\, dN_j/dt + \Sigma h_j\, \Sigma_{\text{out}}\, \dot{n}_j = \xi\Sigma\nu_j h_j. \tag{59}$$

The first term in (59) is identical to the first term in (57). This is also true for the second term since, for any component j, $(\hat{h}_j)_{\text{out}} = (\hat{h}_j)_{\text{in}}$, so that

$$\hat{h}_j\, \Sigma_{\text{out}}\, \dot{n}_j = \hat{h}_j\, [(\dot{n}_j)_{\text{out}} - (\dot{n}_j)_{\text{in}}] = (\hat{h}_j\dot{n}_j)_{\text{out}} - (\hat{h}_j\dot{n}_j)_{\text{in}} = \Sigma_{\text{out}}\, \hat{h}_j\dot{n}_j.$$

The resulting equality between the right-hand sides of (57) and (59) shows, introducing the reaction enthalpy from (51), that

$$\xi\,\Delta\hat{h}_r = \dot{Q} + \dot{W}, \tag{60}$$

which is (56). This generalization is not surprising. Physically speaking, both (54) and (58), with $\nu_j = -1$ for $j =$ fuel, show that ξ is the number of moles of fuel combusted per unit of time.

When several reactions proceed simultaneously with reaction rates ξ_i and reaction enthalpies $\Delta\hat{h}_{ri}$, $i = 1,2,. . .,r$, it follows using (3.80) that (60) can be generalized to

$$\Sigma_r\xi_i\,\Delta\hat{h}_{ri} = \dot{Q} + \dot{W}, \tag{61}$$

or, on a mass basis, using the *specific reaction enthalpy* $\Delta h_{ri} = \Delta\hat{h}_{ri}/\hat{M}_i$ and the *specific reaction rate* $\xi_i' = \xi_i\hat{M}_i$, to,

$$\Sigma_r\,\xi_i'\,\Delta h_{ri} = \dot{Q} + \dot{W}, \tag{62}$$

where the summation Σ_r refers to reactions $i = 1,2,. . .,r$.

The above relations between powers, derived on the basis of the first law applied to the model described by (57), is of importance for the concept of efficiency of growth ((19)) and that of *metabolizable energy*, both introduced in Section 4.3. We note, first, that, under the given conditions, the total energy intake $\dot{E}$ is equal to the sum of accumulated energy $\dot{E}_a$ and maintenance energy $\dot{E}_m$. This intuitive relation expresses an energy balance that can be nothing but the first law. When it is written in the form

$$\dot{E}_a + (-\dot{E}) = (-\dot{E}_m),$$

we see the identity with (57) in that the right-hand side, according to (61), expresses precisely the energy transformations of the chemical reactions associated with maintenance of the metabolism plus the possible contributions that arise from catabolism and synthesis of matter in depots.

The above relation is valid both during the accumulation and deaccumulation of matter (to and from depots), since $\dot{E}_a$ is then positive and negative respectively. In order to calculate efficiencies of accumulation (retention), however, it is necessary to differentiate between *passive* accumulation, i.e. a process where the accumulation takes place without any energy transformations, and *active* accumulation, which requires the participation of other reactions (with negative reaction enthalpies).

Dietary fat can accumulate in depots without any significant coupling to other energy demanding processes. During a period of large fat intake, energy can be accumulated in the form $\hat{h}_i \, dN_i/dt$ (the first term in (57)). However, this increase in the energy content of the system is numerically balanced by the second term in (57) so that the process takes place without any significant change in the sum $(\dot{Q} + \dot{W})$. Accumulation of protein, on the other hand, requires participation of other reactions and cannot take place without increments in the sum $(\dot{Q} + \dot{W})$.

4.5.4 Biological fuels. Indirect calorimetry

In applications of the energy balance for most animals, biological fuels can be divided into three groups: carbohydrate (CH), fat (F) and protein (Pr). Of these, carbohydrate and fat are completely oxidized, while protein is only partly oxidized, with excretion of (mainly) urea in mammals and (mainly) ammonia in fish and marine invertebrates. The physiological heating value of protein is therefore lower (by about 20%) than the value obtained at complete oxidation (e.g. by bomb calorimetry). Also, the chemical composition of the compounds differs within each group of fuels, and so do the molar masses and the molar reaction enthalpies. In spite of these differences, the combustion enthalpies (and the physiological combustion enthalpies) per unit of mass show only a small variation within each group. This is partly due to the relative ratio of the number of atoms that characterize the group and partly to the binding energies of the atoms. It is therefore often helpful to employ mass basis instead of molar basis in formulations of mass and energy conservation.

The reaction enthalpies and stoichiometries of biological fuels shown in Table 4.1 can be used to calculate the energy expenditure of most animals. The respiratory quotient (RQ) denotes the ratio of carbon dioxide elimination and oxygen uptake associated with combustion, and

Table 4.1. *Biological fuels*

Fuel	Specific reaction enthalpy $(kJ/g)^a$ Δh_r	Specific turnover (mmol/g) O_2	(g/g) CO_2	N	RQ	Energy equivalency $(kJ/mol\ O_2)$
CH	−17	33.3	33.3	—	1.00	511
Fat	−39	90.6	63.8	—	0.70	431
Pr	−17	43.3	34.4	0.16	0.79	393

^{a}The fuel value is identical to the negative reaction enthalpy; the value for protein is the physiological value for mammals, where the end product is urea; the value is somewhat higher for aquatic animals, where the end product is ammonia. Carbohydrate represents a mixture of mono-, di- and poly-saccharides.

the last column gives the energy equivalency of oxygen (caloric equivalency). These values are helpful because in many situations it is possible to measure only the rate of uptake of oxygen and the rates of elimination of carbon dioxide and nitrogen. From such values, and knowledge of the reaction enthalpies and stochiometries, it is possible to calculate the energy expenditure.

The energy expenditure is calculated from the first law (Section 4.5.3) with the following assumptions: (i) carbohydrate, fat and protein are the only compounds that undergo combustion; (ii) there is stationarity with respect to all other compounds; and (iii) uptake/elimination of oxygen, carbon dioxide and nitrogen is instantaneous. In this situation, the first law can be written as

$$(\xi'\Delta h_r)_{CH} + (\xi'\Delta h_r)_F + (\xi'\Delta h_r)_{Pr} = \dot{Q} + \dot{W}, \qquad (63)$$

or, taking values from Table 4.1

$$17\,\xi'_{CH} + 39\,\xi'_F + 17\xi'_{Pr} = (-\dot{Q}) + (-\dot{W}) = \dot{E}\ (kW), \qquad (64)$$

where the unit (kW) requires ξ'_i to be given in units of grams of fuel per second.

The conservation of mass is expressed as source terms and is obtained from Table 4.1 as

$$\dot{N}_{(O_2)} = 33.3\,\xi'_{CH} + 90.0\,\xi'_F + 43.3\,\xi'_{Pr}\ (mmol/s), \qquad (65)$$

$$\dot{N}_{(CO_2)} = 33.3\,\xi'_{CH} + 63.8\,\xi'_F + 34.4\,\xi'_{Pr}\ (mmol/s), \qquad (66)$$

$$\dot{M}_N = 0.15\,\xi'_{Pr}\ (g/s) \qquad (67)$$

where, as above, ξ'_i is to be given in units of grams of fuel per second.

Given the experimental values of the quantities on the left-hand sides of (65)–(67), these equations may be solved for the three unknown reaction rates, and the energy expenditure is obtained from (64). This method is called *indirect calorimetry* because the animal, or man, is not actually placed in a calorimeter. However, if the subject is placed in a calorimeter with determination of energy expenditure through measurement of the removed powers $(-\dot{Q})$ and $(-\dot{W})$, we speak of *direct calorimetry*. In practice, any contribution $(-\dot{W})$ from the subject is transformed into heat by way of friction in the calorimeter so that the total energy removed is measured in the form of heat (Atwater and Benedict, 1903; Benedict and Milner, 1907; Jacobsen *et al.*, 1985).

Studies involving the simultaneous measurement of terms on both sides of equation (63) on normal human subjects over periods of up to 72 hours show that the general relation (61) can be reduced to (63). During the first few hours after a meal, however, the left-hand side of (63) is numerically larger than the right-hand side. This result shows that reactions other than those of (63) take place during this time and that at least some of them have positive reaction enthalpies.

Example 4.16

An adult male subject has an oxygen uptake of 21.16 mol over 24 hours and the associated elimination of carbon dioxide and nitrogen is 16.95 mol and 5.76 g, respectively. The subject has performed 0.9 MJ of external work over the same period and his energy expenditure at rest is $\dot{E}_0 = 65$ W. We wish to estimate his energy expenditure, heat loss and net efficiency for the external work.

From (67), and then (65) and (66), we obtain

$$\xi'_{Pr} = 36 \text{ g/d} \; ; \; \xi'_{CH} = 194 \text{ g/d}; \quad \xi'_F = 145 \text{ g/d} \tag{a}$$

so that (64) gives

$$\dot{E} = (-\dot{Q}) + (-\dot{W}) = 9566 \text{ kJ/d} = 110.7 \text{ W}, \tag{b}$$

$$(-\dot{Q}) = 9566 - 900 = 8666 \text{ kJ/d} = 100 \text{ W}. \tag{c}$$

From definition (19) we obtain

$$\eta\text{-net} = (-\dot{W})/(\dot{E} - \dot{E}_0) = 900/(9566 - 5615) = 0.23. \tag{d}$$

The energy equivalency, which expresses the chemical energy that is liberated per mole of oxygen consumed, is different for different fuels (substrates). During combustion of a mixture of carbohydrate, fat and protein with fractions (of the energy of combustion of the food) f_{CH}, f_F and f_{Pr}, the energy equivalency of oxygen is (see Table 4.1)

$$\text{O}_2\text{-equivalency} = 511\, f_{CH} + 431\, f_F + 393\, f_{Pr} \text{ kJ/mol O}_2, \qquad (e)$$

but, as we shall see, it varies only slightly between subjects on a normal diet. Firstly, f_{Pr} varies only slightly around 0.15 so that (e), with $f_{CH} + f_F + f_{Pr} = 1$, can be reduced to

$$\text{O}_2\text{-equivalency} = 493 - 80\, f_F \text{ kJ/mol O}_2. \qquad (f)$$

Secondly, the fraction of fat in the food varies relatively little (between 0.30 and 0.40), so that (f) can be approximated as

$$\text{O}_2\text{-equivalency} \approx 465 \text{ kJ/mol O}_2. \qquad (g)$$

Determination of oxygen consumption alone, assuming a normal diet, will thus generally give rise to small errors only. In the present example, (g) with the observed oxygen consumption, gives an energy expenditure of

$$\dot{E} \approx 465 \times 10^3 \times 21.16/(24 \times 3600) \approx 115 \text{ W}, \qquad (h)$$

which deviates less than 4% from the result (b).

Example 4.17

An amphipod with a body weight of 10 μg consumes 4.0×10^{-9} mol oxygen every hour at steady state and eliminates 3.6×10^{-9} mol carbon dioxide, 0.4×10^{-a} mol N (as ammonia) and 0.1×10^{-9} mol lactic acid. The external work power is 50×10^{-9} W. We wish to calculate the heat loss of the animal when the following four net reactions contribute to the energy expenditure,

$$C_6H_{12}O_6 + 6\,O_2 \rightarrow 6\,CO_2 + H_2O \quad \{+2870\}, \qquad (a)$$

$$C_6H_{12}O_6 \rightarrow 2\,C_3H_6O_3 \quad \{+\ 100\}, \qquad (b)$$

$$C_{55}H_{104}O_6 + 78\,O_2 \rightarrow 55\,CO_2 + 52\,H_2O \quad \{+34\,300\}, \qquad (c)$$

$$C_{32}H_{48}O_{10}N_8 + 33\,O_2 \rightarrow 32\,CO_2 + 8\,NH_3 + 12\,H_2O \quad \{+14\,744\}, \qquad (d)$$

where $\{-\Delta\hat{h}_r\}$ in kJ/mol substrate is given for each reaction. Note that the real protein molecule can be regarded as a macromolecule consisting of about 100 molecules with chemical composition as given in (d), so that the molar mass is about 77 600 g/mol. The calculations follow those of Example 4.16, but on a molar basis, since reaction stoichiometries are given. The energy expenditure is calculated from (61), rewritten with the given reaction enthalpies as

$$\dot{E} = 2870\ \xi_G + 100\ \xi_{G-La} + 34\ 300\ \xi_F + 14\ 744\ \xi_{Pr}, \qquad (e)$$

where the molar reaction rates refer to (a)–(d), with G (combustion of glucose), G–La (glucose metabolism to lactic acid, anaerobic process), F (combustion of fat) and Pr (combustion of protein). The source terms for mass conservation are, in accordance with (a)–(d),

$$\dot{N}(O_2) = 6\ \xi_G + 78\ \xi_F + 33\ \xi_{Pr} = 4 \times 10^{-9}\ \text{mol/h}, \qquad (f)$$

$$\dot{n}(La) = 2\ \xi_{G-La} = 0.1 \times 10^{-9}\ \text{mol/h}, \qquad (g)$$

$$\dot{N}(CO_2) = 6\ \xi_G + 55\ \xi_F + 32\ \xi_{Pr} = 3.6 \times 10^{-9}\ \text{mol/h}, \qquad (h)$$

$$\dot{N}(N) = 8\ \xi_{Pr} = 0.4 \times 10^{-9}\ \text{mol/h}. \qquad (i)$$

Solving (f)–(i) for reaction rates and substituting them into (e) gives

$$\dot{E} = (2870 \times 0.194 + 100 \times 0.05 + 34\ 300 \times 0.0152$$
$$+ 14\ 744 \times 0.05) \times 10^{-9} = 1820 \times 10^{-9}\ \text{kJ/h} = 0.506\ \mu\text{W}, \quad (j)$$

and the heat loss becomes

$$(-\dot{Q}) = \dot{E} - (-\dot{W}) = 0.506 - 0.050 = 0.46\ \mu\text{W}. \qquad (k)$$

Example 4.18

The data for the balanced reaction equation (d) in Example 4.17 for combustion of 'standard' protein can be recalculated to produce data which conform with those of Table 4.1. The monomer of the macromolecule has a molar mass of 776 g/mol and a molar reaction enthalpy of $-14\ 744$ kJ/mol:

$$\Delta \hat{h}_r = -14\ 744/776 = -19.0\ \text{kJ/g}, \qquad (a)$$

$$\nu(O_2)/\hat{M}(Pr) = (33/776) \times 1000 = 42.5\ \text{mmol/g}, \qquad (b)$$

$$\nu(CO_2)/\hat{M}(Pr) = (32/776) \times 1000 = 41.2\ \text{mmol/g}, \qquad (c)$$

$$M(N)/M(Pr) = 8 \times 14/776 = 0.14\ \text{g/g}, \qquad (d)$$

$$RQ = \dot{V}(CO_2)/\dot{V}(O_2) = N(CO_2)/N(O_2) = 32/33 = 0.97, \quad \text{(e)}$$

$$-\Delta\hat{h}_r/\nu(O_2) = 14\,744/33 = 447\,\text{kJ/mol } O_2. \quad \text{(f)}$$

The protein is from an aquatic animal and the numbers deviate to some extent from those in Table 4.1.

4.5.5 Coupled chemical reactions

The first law for processes in control volumes in which there are several simultaneous chemical reactions was formulated in Section 4.5.3. With a few assumptions that are quite realistic for many biological systems (isothermal, isobaric process and stationarity with respect to intermediary products), the result could be generalized to (62), which for two stationary or non-stationary processes can be written as

$$\xi_1\Delta\hat{h}_{r1} + \xi_2\,\Delta\hat{h}_{r2} = \dot{Q} + \dot{W} \text{ (CV)}. \quad (68)$$

From an energy point of view, some reactions are *exothermic* and liberate energy ($\Delta\hat{h}_r < 0$), as, for example, the combustion of glucose (46), while other reactions are *endothermic* and consume energy ($\Delta\hat{h}_r > 0$), as, for example, the synthesis of ATP,

$$\text{ADP} + \text{phosphate} \rightarrow \text{ATP} + H_2O. \quad (69)$$

The first law contains no restrictions on the direction of reactions and they can proceed alone or simultaneously. However, the second law must also be fulfilled and it shows (Section 5.4.2) that only some reactions can proceed spontaneously and alone, such as (46), while other reactions, such as (69) or the reverse of (46), cannot. However, the latter can proceed with help from other processes, in particular from other chemical reactions. This assistance is not a supply of energy as heat, as is the case for some endothermic processes (e.g. reduction of some oxides), but, rather, a supply of chemical energy of 'high quality'. It can be shown that both the first and second laws can be fulfilled when the two reactions (46) and (69) proceed simultaneously, as shown in schematic form in Figure 4.14(*a*).

When several simultaneous reactions proceed under the conditions described above, we speak of *coupled reactions*. When such reactions are analysed in energy terms, it is useful to consider the reactions separately and to calculate the necessary transfer of energy. This is shown in Figure

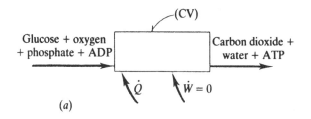

(a)

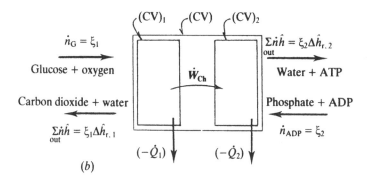

(b)

Fig 4.14 Isotherm process with two coupled reactions: (a) total process; (b) two separate reactions with transfer of internal energy.

4.14(b) for the present example by a curved arrow for the *flow of internal energy* $\dot{W}_{Ch}$. This internal energy transfer (invisible from the outside) should be understood as useful energy (availability, Gibbs energy or chemical work, Section 3.5.4). It is supplied by $(CV)_1$ and consumed by $(CV)_2$. As an example, the transfer could take place by way of free energy associated with matter since a product from one reaction could be a reactant in another reaction. Also, a number of intermediary reactions may be involved (see Blaxter, 1989), even though only the net result of reactions (46) and (69) can be observed. In any case, the first law can be formulated for each of two control volumes as

$$\xi_1\, \Delta \hat{h}_{r1} = \dot{Q}_1 - \dot{W}_{Ch}\ (CV)_1, \tag{70}$$

$$\xi_2\, \Delta \hat{h}_{r2} = \dot{Q}_2 + \dot{W}_{Ch}\ (CV)_2. \tag{71}$$

The sum of (70) and (71) gives (68), which applies to the total control volume (CV) when $\dot{W} = 0$ and $\dot{Q} = \dot{Q}_1 + \dot{Q}_2$.

4.5.6 Photosynthesis

It is well known that different photochemical reactions have played a central role in the formation of the first organic molecules. Since that time, photosynthesis in green plants, as we know it today, has taken over the important role of producing, as the only photochemical reaction, the organic material that is necessary for all life. This includes plants and animals in view of the constant breakdown or organic matter in the ecological cycle. In addition, necessary production of oxygen is maintained.

The reverse of reaction (46), the combustion of glucose, is the net reaction for photosynthesis, with the formation of glucose (hexose) and oxygen from carbon dioxide and water:

$$6\ CO_2 + 6\ H_2O \rightarrow C_6H_{12}O_6 + 6\ O_2. \tag{72}$$

The enthalpy of reaction of (72) is positive, $\Delta \hat{h}_r = 478\ kJ$ per mol CO_2 at 25 °C. The reaction requires a supply of energy.

Process (72) proceeds in green plants and is catalysed by molecules of chlorophyll, which absorb radiant energy in a relatively narrow range of wave lengths in the visible part of the electromagnetic spectrum (for chlorophyll A about 680 nm). The resulting transformation of electromagnetic energy into chemical energy amounts to only a very small part (1–3%) of the incident radiation. The remaining part is absorbed as heat or is reflected. The plant is therefore heated until it reaches a stationary temperature at which the heat losses due to convection and radiation balance the heat absorbed.

Reaction (72) is one of several reactions by which the plant can synthesize organic matter such as glucose, starch, cellulose, fat, etc. Concomitantly, oxygen is used in all combustion reactions.

Example 4.19

The intensity of the sun's radiation on a clear day is observed to be $\dot{Q}_S/A = 50\ mW/cm^2$. Calculate the accumulation of glucose during eight hours of exposure in a green leaf. Its surface area is $10\ cm^2$ and the heat loss from the leaf is $\dot{Q}_1/A = 49\ mW/cm^2$. It is assumed that the temperature of the leaf is steady at 25 °C and that only the net process (72) proceeds (reaction enthalpy $\Delta \hat{h}_r = 478\ kJ/mol\ CO_2$ at 25 °C).

If ξ is the reaction rate in moles of CO_2 consumed per second, the mole balance ((58)) for glucose (G), using (72), can be written as

$$dN_G/dt = \xi/6, \tag{a}$$

and the first law ((61)) for a single reaction as

$$\xi \, \Delta\hat{h}_r = \dot{Q}_s - \dot{Q}_1. \tag{b}$$

Note that the conditions for (61) are fulfilled regardless of whether reactants and products accumulate or deaccumulate in the leaf, as long as there is stationarity with respect to intermediates. Accumulation of glucose is described by (a), and (a) and (b) give

$$dN_G/dt = \tfrac{1}{6}\,\xi = \tfrac{1}{6}\,(\dot{Q}_s - \dot{Q}_1)/\Delta\hat{h}_r. \tag{c}$$

Over a period of eight hours this becomes

$$N_G = 8 \times 60 \times 60 \times (\tfrac{1}{6}) \times (50 - 49) \times 10 \times 10^{-3}/478 = 0.1 \text{ mmol.} \tag{d}$$

In accordance with stoichiometry ((72)), 0.6 mmol of CO_2 will be fixed in the leaf over the eight-hour period.

Example 4.20

Calculate the number of photons of wave length $\lambda = 680$ nm required to fix one molecule of CO_2 in reaction (72). For $h = 6.626 \times 10^{-34}$ Js (the Planck constant), $c = 3 \times 10^8$ m/s (the velocity of light) and $\nu = c/\lambda$ (the frequency), the energy per photon is

$$h\nu = 6.626 \times 10^{-34} \times 3 \times 10^8/(680 \times 10^{-9}) = 2.9 \times 10^{-19} \text{ J}, \tag{a}$$

and the reaction enthalpy per molecule of CO_2 is

$$\Delta\hat{h}_r/N_A = 478 \times 10^3/6.022 \times 10^{23} = 7.94 \times 10^{-19} \text{ J}. \tag{b}$$

The ratio of (b) and (a) shows that energy from about three photons is required to fix one molecule of CO_2, or about $6 \times 3 = 18$ photons in order to produce one molecule of glucose. Experiments using radiation of monochromatic light at 680 nm show quantum efficiencies of up to 30%, hence plants can use about one-third of all the incident photons. In Example 4.19, utilization of $\tfrac{1}{50}$, or 2%, of total incident light was assumed. This utilization is typical and shows that absorption with concomitant photosynthesis is a quantum effect associated with a very narrow range of wave lengths.

4.6 Electrochemical energy transformation

In this section, we deal with processes in which electrically charged components are subjected to movement from one electrical potential to another. Such processes take place in living systems across membranes around and within cells and are often coupled to chemical reactions.

As described later (Chapter 7), the mass flux in an isothermal diffusion process is determined by a difference in electrochemical potential. This is analogous to the heat flux for the transport of heat energy in, e.g., solid matter where the transport is determined by a difference in temperature. The direction of spontaneous processes is given by the second law. The required energy supply is given by the first law.

In living systems, mass fluxes are often observed to proceed against the electrochemical potential gradient. The necessary Gibbs energy required is then supplied by a chemical reaction (chemical pump) that is coupled to the transport process and determines the rate of the associated charge flux. Such a transport is called an *active transport* and, in addition to the supply of Gibbs energy, it also requires a supply of energy. It should be noted that spontaneous transport processes can also receive a supply of energy and Gibbs energy whereby the mass flux is augmented.

If changes in kinetic energy are neglected and if only electric potential energy is included in (9), the first law for a control volume can be written in the compact form

$$d(N\bar{u})/dt + \Sigma_{\text{out}} \, \dot{n} \, \bar{h} = \dot{Q} + \dot{W} \text{ (CV)}, \qquad (73)$$

where $\bar{u} = \hat{u} + \hat{\varphi}$, $\bar{h} = \hat{h} + \hat{\varphi}$, according to (3.133) and (3.134), and $\hat{\varphi} = z\mathscr{F}\mathscr{E}$ according to (3.18), where z is the valency of the ion, $\mathscr{F}$ the Faraday constant, and $\mathscr{E}$ the electrical potential. For mixtures, $N\bar{u}$ and $\dot{n}\bar{h}$ are the sums of the contributions from all components.

To exemplify a process involving active transport, we now consider the system in Figure 4.15, which shows the coupled reactions separately (cp. Section 4.5.5). The diagram illustrates the fundamental process for active transport of Na^+ ions from the lower electrical potential $\mathscr{E}_i$ in the intracellular space to the higher potential $\mathscr{E}_e$ in the extracellular space.

The net reactions for energy transformation begin with the combustion of glucose ((46)), which supplies chemical energy $\dot{W}_{\text{Ch1}}$ to synthesis of ATP (69). The energy containing ATP, in the hydrolysis

$$ATP + H_2O \rightarrow ADP + \text{phosphate}, \qquad (74)$$

can supply chemical energy W_{Ch2} to processes (not shown here) that are coupled to the transport ξ_3 of Na^+-ions to the extracellular space. The return flux ξ_3' of Na^+-ions takes place as a passive diffusion. The heat powers $(-\dot{Q}_1)$, $(-\dot{Q}_2)$, etc., are expected to be unavoidable consequences of the second law.

For the following analysis, and for more general considerations, the system in Figure 4.15 can be considered to contain five processes (reactions). The vertical arrows symbolize reaction rates ξ_i (moles per unit time); $\xi_1 = \dot{n}_G$ for glucose combustion, $\xi_2' = \dot{n}_{ATP}$ hydrolysis and ξ_2 for the reverse synthesis, $\xi_3' = \dot{n}_{Na}$ for leakage of sodium ions and $\xi_3 = \dot{n}_{Na}$ for the active transport.

A detailed analysis of the processes in Figure 4.15 requires the choice of several control volumes, i.e. the five shown in the diagram. We will assume stationarity of all processes for now. Instationarity, which would imply accumulation/deaccumulation of ATP, ADP and Na^+-ions, will be included in the general model of a biological system discussed in Section 4.7. Proper choice of control volumes in Figure 4.15 and use of mass and energy conservation implies, first, that

$$\xi_2' = \xi_2; \quad \xi_3' = \xi_3; \quad \dot{W}_{Ch1} = \dot{W}_{Ch2} - \dot{Q}_2 - \dot{Q}_2'. \tag{75}$$

Then, the first law for each of the five control volumes gives

$$\xi_1 \, \Delta\hat{h}_{r,G} = \dot{Q} - \dot{W}_{Ch1} \ (CV)_1, \tag{76}$$

$$\xi_2(-\Delta\hat{h}_{r,ATP}) = \dot{Q}_2 + \dot{W}_{Ch1} \ (CV)_2 \tag{77}$$

$$\xi_2' \, \Delta\hat{h}_{r,ATP} = \dot{Q}_2' - \dot{W}_{Ch2} \ (CV)_2', \tag{78}$$

$$\xi_3 \, z_{Na^+} \, \mathscr{F} \, (\mathscr{E}_e - \mathscr{E}_i) = \dot{Q}_3 + \dot{W}_{Ch2} \ (CV)_3, \tag{79}$$

$$\xi_3' \, z_{Na^+} \, \mathscr{F} \, (\mathscr{E}_i - \mathscr{E}_e) = \dot{Q}_3' \ (CV)_3', \tag{80}$$

where (76) is based on (60) and where (79)–(80) is based on the assumptions of isothermal and ideal mixtures so that the molar enthalpy of Na^+ ions is independent of the concentrations c_e and c_i. Clearly, the first law for the sum of two control volumes is the sum of the first law for the two. Using (75), the sum of (76)–(80) becomes

$$\xi_1 \, \Delta\hat{h}_{r,G} = \dot{Q}_1 + \dot{Q}_2 + \dot{Q}_2' + \dot{Q}_3 + \dot{Q}_3' \ (CV). \tag{81}$$

Each side of (81) is also the negative of the energy expenditure of CV, cp. (10) and (11).

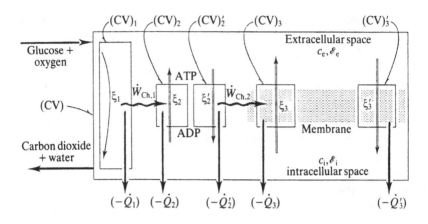

Fig 4.15 Active transport of Na^+ ions from intracellular space to extracellular space in the direction of an increasing electrochemical potential.

The sum of (76)–(79), with $\xi_1 = \dot{n}_G$ and $\xi_3 = \xi_3' = \dot{n}_{Na+}$, gives

$$\dot{n}_G \, \Delta\hat{h}_{r,G} + \dot{n}_{Na+} \, z_{Nat} \mathscr{F} \, (\mathscr{E}_e - \mathscr{E}_i) = \dot{Q}_1 + \dot{Q}_2 + \dot{Q}_2' + \dot{Q}_3, \qquad (82)$$

which shows that part of the chemical energy liberated in the combustion of glucose supplies the flow of electrical potential energy and that part is removed as heat. On this basis we may define an efficiency ((14)) for the transformation of chemical energy into electrical energy:

$$\eta = [\dot{n}_{Na+} \, z_{N_{Na+}} \mathscr{F} \, (\mathscr{E}_e - \mathscr{E}_i)]/(-\dot{n}_G \, \Delta\hat{h}_{r,G}). \qquad (83)$$

An efficiency for the transformation of chemical exergy into electrical exergy will be discussed in Section 8.3.

In conclusion, the active transport of Na^+ ions from a lower to a higher concentration and from a lower to a higher electrical potential requires, as expected, a supply of energy $\dot{W}_{Ch2}$, cp. (79). However, the magnitude of this energy supply is neither determined by the change in electro-chemical enthalpy fluxes alone (which would be zero for the transport of a neutral component between ideal mixtures), nor by the first law alone. The explanation of this apparent paradox is to be found in the second law, which imposes requirements on the magnitude of $\dot{Q}_3$ (Section 5.5), and in the exergy balance, which imposes explicit requirements with respect to the minimum supply of chemical energy $\dot{W}_{Ch2}$ (Section 8.4.2).

Example 4.21

We consider a tissue in which Na^+-ions are transported actively from the intracellular to the extracellular space. The transport takes place through participation of hydrolysis of ATP and it is assumed that 40% of the energy of hydrolysis can be used for the transport. It is, furthermore, assumed that the energy required for the formation of ATP results from the combustion of glucose (G), and that 38 mol of ATP are formed per mol of glucose combusted. In the steady state, the extracellular concentration of Na^+ is observed to be 145 mM and the intracellular concentration to be 12 mM; and the electrical potential of the intracellular fluid is determined to be -90 mV (in relation to the extracellular space). The transcellular flux of Na^+ ions is 1 mmol per minute and the temperature of the tissue is 37 °C.

We wish to calculate the oxygen consumption, heat output and energy expenditure of the process when $\Delta \hat{h}_r = -20$ kJ/mol for hydrolysis of ATP and $\Delta \hat{h}_r = -2867$ kJ/mol for glucose combustion. The solution of this detailed example requires information about the partition of the total heat power ((81)) among the five different parts of the total process. The unknowns in the example are seen to be two reaction rates (ξ_1 and ξ_2), two internal energy flows and five heat powers, while there are, besides (76)–(80), only two pieces of additional information (expressed by the two first conditions in (a) below). The two remaining conditions are now fulfilled by the assumptions that $\dot{Q}_2/\dot{Q}_1 = 1$ and $\dot{Q}_3/\dot{Q}_3' = \frac{4}{5}$, to be discussed in more detail in Example 5.12.

The processes are described schematically in Figure 4.15 and the first law for the control volumes depicted is given by (76)–(81). The information given above leads to

$$\dot{W}_{Ch2} = 0.4\, \dot{W}_{Ch1}; \quad \xi_2/\xi_1 = 38; \quad \dot{Q}_2/\dot{Q}_1 = 1\ ; \quad \dot{Q}_3/\dot{Q}_3' = \frac{4}{5}. \quad \text{(a)}$$

Then (80) gives

$$-Q_3' = (0.001/60) \times (1) \times 96\,500 \times 0.090 = 0.145\ \text{W}, \quad \text{(b)}$$

which, using (a), in particular $\dot{Q}_3 = (\frac{4}{5})\, \dot{Q}_3'$, in (79) gives

$$\dot{W}_{Ch2} = (9/5)\, \dot{Q}_3' = 0.261\ \text{W}, \ \dot{W}_{Ch1} = 0.261/0.4 = 0.653\ \text{W}. \quad \text{(c)}$$

Using the second and third assumptions in (a), $\dot{Q}_1$ is eliminated from (76) and (77) and the result can be solved for the reaction rates:

$$\dot{n}_G = \dot{\xi}_1 = -2\dot{W}_{Ch1}/(\Delta\hat{h}_{r,G} + 38\,\Delta\hat{h}_{r,ATP})$$
$$= 2 \times 0.653/(-2867 - 38 \times 20) = 0.360\,\mu\text{mol/s}, \tag{d}$$

$$\dot{n}_{ATP} = \dot{\xi}_2 = 38 \times 0.360 = 13.7\,\mu\text{mol/s}. \tag{e}$$

The oxygen consumption is obtained from (46):

$$\dot{n}(O_2) = 6\,\dot{n}_G = 2.16\,\mu\text{mol/s}. \tag{f}$$

Furthermore, the heat powers are calculated from (76), (a) and (78)

$$-\dot{Q}_1 = 0.360 \times 10^{-3} \times 2867 = 0.379\,\text{W}, \tag{g}$$

$$-\dot{Q}_2 = 0.379\,\text{W}, \tag{h}$$

$$-\dot{Q}_2' = 13.7 \times 10^{-3} \times 20 - 0.261 = 0.013\,\text{W}, \tag{i}$$

$$-\dot{Q}_3' = (4/5) \times 0.145 = 0.116\,\text{W}, \tag{j}$$

while $\dot{Q}_3' = 0.145\,\text{W}$, as obtained above. The total energy expenditure becomes, according to (11) and (81),

$$\dot{E} = 0.397 + 0.397 + 0.013 + 0.116 + 0.145 = 1.032\,\text{W}. \tag{k}$$

The energy efficiency (83) becomes

$$\eta = 0.145/1.032 = 0.14. \tag{l}$$

Note that the Na^+ ion concentrations in the extra- and intracellular spaces are not used in the calculations because of the assumptions of isobaric, isothermal and ideal mixtures. However, it is possible, by considering some typical values of enthalpies of mixing, to estimate the errors introduced by the assumption of ideal mixtures. Thus, the molar enthalpy change of isobaric and isothermal dilution to infinity of a 0.1 M solution of NaCl is -348 J/mol at 25 °C and the molar enthalpy change of dilution from 0.1 M to 0.0125 M is -180 J/mol (Harned and Owen, 1958). From these values, the change in enthalpy of mixing for Na^+ ions in the process in $(CV)_3$ in Figure 4.15 can be estimated to be $\hat{h}_e - \hat{h}_i < 200$ J/mol Na^+. This contribution should be included in (79) which, for non-ideal solutions, then takes the form

$$\dot{\xi}_3[(\hat{h}_e - \hat{h}_i) + z(Na^+)\mathscr{F}(\mathscr{E}_e - \mathscr{E}_i)] = \dot{Q}_3 + \dot{W}_{Ch2}. \tag{m}$$

The ratio between the first and second terms within square brackets in (m) is

$$200/(1 \times 96\ 500 \times 0.09) \approx 0.02, \tag{n}$$

so that the error is less than 2%.

4.7 Model of a biological system

In the preceding section, the analysis of energy transformations associated with combustion was performed without any explicit discussion of the internal processes that proceed simultaneously; the assumption was made that such processes proceed in a steady fashion. Determination of the consumption of combustible matter (e.g. indirectly by way of transformations of oxygen, carbon dioxide and nitrogen) made it possible to determine the energy expenditure of the living system.

In this section, we formulate and analyse a simplified model of a living system containing a number of fundamental internal processes. The analysis shows how to account for the chemical energy liberated in the combustion reactions before this energy is ultimately removed from the system as heat – apart from that amount used to perform external work.

The basal processes shown in Figure 1.6 are now quantified as shown in Figure 4.16. They are as follows:

1. The fundamental process of energy supply in which fuel (A) is combined with oxygen (B) to yield products (C) and (D), which are removed from the system. This process takes place locally throughout the system and stepwise through stationary intermediate reactions (cp. Figure 4.12), which capture liberated energy by the transformation of ADP to 'energy-rich' ATP. The reverse process can liberate and transfer energy $\dot{W}_{\mathrm{Ch}i}$ to other fundamental processes. These processes are exemplified by the following.
2. Synthesis of product (AA), e.g. glycogen, from fuel (A), e.g. glucose.
3. Synthesis of product (FF) from absorbed reactant (F).
4. Internal pumps which maintain local differences in electrochemical energy and so 'lift' a non-structure (L')into a structure (L"), e.g. a low intracellular concentration of Na^+ ions into a higher extracellular one.
5. Creation of muscle tension for the performance of internal and external work.

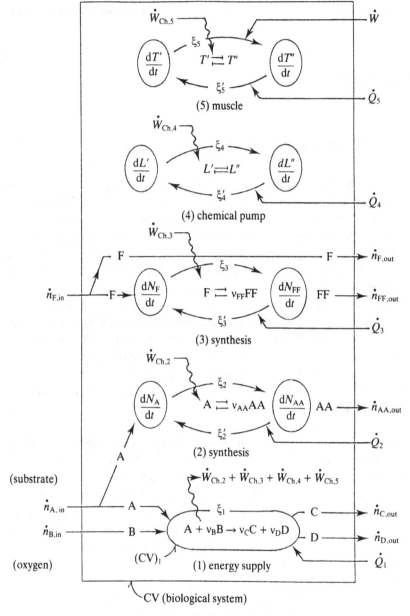

Fig 4.16 Model of a biological system. (1) supply of energy for use in four basic processes; (2) synthesis; (3) synthesis; (4) chemical pump; and (5) muscle which delivers internal and external work. Only external work $\dot{W}$ is shown: $d(\ )/dt$ symbolizes depots for accumulation/deaccumulation of matter, tension and energy.

Table 4.2. *Model of the biological system illustrated in Figure 4.16*

Process	Rate	Process enthalpy
1 (energy supply)	ξ_1	$\Delta\hat{h}_{r1} = \nu_C\hat{h}_C + \nu_D\hat{h}_D - \hat{h}_A - \nu_B\hat{h}_B$
2 (synthesis)	ξ_2	$\Delta\hat{h}_{r2} = \nu_{AA}\hat{h}_{AA} - \hat{h}_A$
2' (decomposition)	ξ_2'	$\Delta\hat{h}_{r2}' = -\Delta\hat{h}_{r2}$
3 (synthesis)	ξ_3	$\Delta\hat{h}_{r3} = \nu_{FF}\hat{h}_{FF} - \hat{h}_F$
3' (decomposition)	ξ_3'	$\Delta\hat{h}_{r3}' = -\Delta\hat{h}_{r3}$
4 (chemical pump)	ξ_4	$\Delta\hat{h}_{r4} = \nu_{L''}\hat{h}_{L''} - \hat{h}_{L'}$
4' (back diffusion)	ξ_4'	$\Delta h_{r4}' = -\Delta\hat{h}_{r4}$
5 (muscle tension)	ξ_5	$\Delta\hat{h}_{r5} = \hat{h}_{T''} - \hat{h}_{T'}$
5' (relaxation)	ξ_5'	$\Delta\hat{h}_{r5}' = -\Delta\hat{h}_{r5}$

The last four structure-building processes proceed with forward reaction rates ξ_2, ξ_3, ξ_4 and ξ_5 and with reverse structure-decomposing rates ξ_2', ξ_3', ξ_4' and ξ_5'. For example, ξ_2 is the consumption of A in mol/s and ξ_2' the production of A in mol/s by the reverse process (2). Thus, steady conditions imply $\xi_2 = \xi_2'$. There is a continuous excretion of AA and FF from the system, and there is, in general, accumulation or deaccumulation in the internal depots dN_j/dt as well as in the depots of structure (L') and (L'') and mechanical energy (T') and (T''). Each of the processes is associated with release of energy as heat, symbolized in Figure 4.15 as being associated with the reverse processes. The model explicitly includes the performance of external work in process (5).

The conceptual biological model is thus described by the nine processes shown in Table 4.2, along with their process rates and enthalpies.

Energy conservation for the whole system, i.e. (CV) in Figure 4.16, satisfying the conditions implicit in (61), can now be written as

$$\xi_1\Delta\hat{h}_{r1} + (\xi_2 - \xi_2')\,\Delta\hat{h}_{r2} + (\xi_3 - \xi_3')\,\Delta\hat{h}_{r3} + (\xi_4 - \xi_4')\,\Delta\hat{h}_{r4} \\ + (\xi_5 - \xi_5')\,\Delta\hat{h}_{r5} = \dot{Q} + \dot{W}, \tag{84}$$

where the total added heat power can be identified from the diagram as

$$\dot{Q} = \dot{Q}_1 + \dot{Q}_2 + \dot{Q}_3 + \dot{Q}_4 + \dot{Q}_5, \tag{85}$$

and the energy expenditure, according to (11), becomes

$$\dot{E} = (-\dot{Q}) + (-\dot{W}). \tag{86}$$

Since the different parts of the whole process are known, each one can be analysed by defining an appropriate control volume and then writing mass and energy conservation for it, as we now demonstrate.

The molar fluxes in processes 1 and 2 are coupled so that the mass conservation for component A must be written for $(CV)_{1+2}$ (control surface not shown):

$$dN_A/dt - \dot{n}_{A,in} = -\xi_1 - (\xi_2 - \xi_2').\tag{87}$$

The remaining mole balances for processes 1 and 2 can be written separately for $(CV)_1$:

$$-\dot{n}_{B,in} = -\nu_B\xi_1; \quad \dot{n}_{C,out} = \nu_C\xi_1; \quad \dot{n}_{D,out} = \nu_D\xi_1,\tag{88}$$

and for $(CV)_2$, which includes reaction and depots:

$$dN_{AA}/dt + \dot{n}_{AA,out} = \nu_{AA}(\xi_2 - \xi_2').\tag{89}$$

The energy balance ((61)) can also be written for each of the control volumes, with the internal energy transfer $\dot{W}_{Chi}$ being included. For $(CV)_1$, we obtain

$$\xi_1 \Delta\hat{h}_{r2} = \dot{Q}_1 - (\dot{W}_{Ch2} + \dot{W}_{Ch3} + \dot{W}_{Ch4} + \dot{W}_{Ch5}),\tag{90}$$

where the negative of each side represents the energy expenditure. For $(CV)_2$, we obtain

$$(\xi_2 - \xi_2') \Delta\hat{h}_{r2} = \dot{Q}_2 + \dot{W}_{Ch2}.\tag{91}$$

Process 3 can be analysed separately for the conservation of mass of components F and FF respectively. $(CV)_3$ for this process, including the depots, gives

$$dN_F/dt + \dot{n}_{F,out} - \dot{n}_{F,in} = -(\xi_3 - \xi_3'),\tag{92}$$

$$dN_{FF}/dt + \dot{n}_{FF,out} = \nu_{FF}(\xi_3 - \xi_3').\tag{93}$$

Energy conservation for the mixture gives

$$(\xi_3 - \xi_3') \Delta\hat{h}_{r3} = \dot{Q}_3 + \dot{W}_{Ch3}.\tag{94}$$

In a similar way, we obtain $(CV)_4$

$$dL'/dt = -dL''/dt = -(\xi_4 - \xi_4'),\tag{95}$$

$$(\xi_4 - \xi_4') \Delta\hat{h}_{r4} = \dot{Q}_4 + \dot{W}_{Ch4},\tag{96}$$

and, for $(CV)_5$

$$dT'/dt = - dT''/dt = - (\xi_5 - \xi_5'), \tag{97}$$

$$(\xi_5 - \xi_5') \, \Delta\hat{h}_{r5} = \dot{Q}_5 + \dot{W} + \dot{W}_{Ch5}. \tag{98}$$

Inspection of (91), (94), (96) and (98) shows that $\dot{W}_{Ch2}$, $\dot{W}_{Ch3}$, $\dot{W}_{Ch4}$ and $\dot{W}_{Ch5}$ are the energy expenditure of the respective internal control volumes. For example, the energy expenditure of $(CV)_5$, which performs the external work, is

$$\dot{W}_{Ch5} = (-\dot{Q}_5) + (-\dot{W}) + (\xi_5 - \xi_5') \, \Delta\hat{h}_{r5}, \tag{99}$$

where, as expected, $\Delta\hat{h}_{r5} > 0$, $\xi_5 > \xi_5'$ and $(-\dot{W}) > 0$.

For the whole system, i.e. (CV), we can use (57) to identify three contributions to the energy balance of interest to physiologists and nutritionists: *passive energy accumulation (deaccumulation), energy intake* and *energy outflow*. They are given as the sums $\Sigma\hat{h}_i \, dN_i/dt$, $\Sigma\hat{h}_i \, \dot{n}_{i,\text{in}}$ and $\Sigma\hat{h}_i \, \dot{n}_{i,\text{out}}$, respectively, with the simplification that summations only include compounds that are part of processes associated with energy changes within the body.

However, physiologists and nutritionists traditionally define these three contributions differently, in effect such that they are calculated using combustion enthalpies of combustible substrates only. Thus, energy accumulation (deaccumulation) has been defined as $\Sigma \, \Delta\hat{h}_{j,c} \, dn_j/dt$, where $\Delta\hat{h}_{j,c}$ stands for the enthalpy of the combustion of substrate j. Likewise, energy intake and energy outflow have been defined as $\Sigma \, \Delta\hat{h}_{j,c} \, \dot{n}_{j,\text{in}}$ and $\Sigma \, \Delta\hat{h}_{j,c} \, \dot{n}_{j,\text{out}}$, respectively, involving only the substrates flowing into and out of the body.

The magnitude of the three contributions is usually different in the two representations, but taken together, according to an energy balance, the three contributions must amount to the same total energy change for the system, provided the term substrate includes *all* the reacting components. Recall (cp. Section 4.5.2) that the enthalpy of combustion of substrate j, $\Delta\hat{h}_{j,c}$, equals the negative of the reaction enthalpy using the proper definition of reference states.

> For the steady reaction in Example 4.14, there is zero intake and an outflow of negative energy, according to (57). For a total energy balance, this is of course equivalent to a positive intake of energy (combustion enthalpy of substrate H_2) and zero outflow, according to the alternative view. This view is also intuitively clear for an unsteady process, say in which fuel H_2 is accumulated in the control volume of Figure 4.11. Here, the intake of combustion enthalpy with inflow of H_2

is a positive quantity that equals the amount accumulated. According to (57), on the other hand, both enthalpy inflow and energy accumulated are zero. But the accumulated mass of H_2 has been registered by the conservation of mass, and it holds a potential chemical energy content that may become useful in a later deaccumulation and combustion process in which outflow of negative enthalpy then reveals the release of energy. Later (in Chapter 8) we will see that a balance of useful energy (exergy) conceptually follows the traditional view.

Below, in further discussion of the specific processes of the model of a biological system, we confine our attention to the first representation, based on (57). Furthermore, we combine the second and third contributions and define two terms of physiological interest: *energy accumulation (deaccumulation) and net energy intake*. These terms are used to describe the situation over time periods of several days for man (shorter times for smaller organism), and numerical values of these terms represent average values. In addition to the assumptions implicit in (57), we then make the reasonable assumptions (i) that $dN_j/dt = 0$ for components B, C and D; and (ii) that $dT'/dt = dT''/dt = 0$ and $dL'/dt = dL''/dt = 0$. In this case, *the rate of increase in energy content* of the system is

$$\dot{E}_a = \hat{h}_F dN_F/dt + \hat{h}_{FF} dN_{FF}/dt + \hat{h}_A\, dN_A/dt + \hat{h}_{AA}\, dN_{AA}/dt$$
$$= \Sigma \hat{h}_i\, dN_i/dt, \tag{100}$$

and the *net energy intake* is

$$\dot{E}_i = -[\hat{h}_F\, \dot{n}_F\, \dot{n}_{F,\text{out}} + \hat{h}_{FF}\, \dot{n}_{FF,\text{out}} + \hat{h}_{AA}\, \dot{n}_{AA,\text{out}}$$
$$+ \hat{h}_C\, \dot{n}_{C,\text{out}} + \hat{h}_D\, \dot{n}_{D,\text{out}}$$
$$- \hat{h}_F\, \dot{n}_{F,\text{in}} - \hat{h}_A\, \dot{n}_{A,\text{in}} - \hat{h}_B\, \dot{n}_{B,\text{in}}] = -\Sigma_{\text{net out}}\, \dot{n}_i \hat{h}_i. \tag{101}$$

Now, using the assumption (i) above, i.e. stationarity with respect to B, C and D (oxygen, carbon dioxide and water), (101) can be written as

$$\dot{E}_i = -[\hat{h}_F\, (\dot{n}_{F,\text{out}} - \dot{n}_{F,\text{in}}) + \hat{h}_{FF}\, \dot{n}_{FF,\text{out}}$$
$$+ \hat{h}_{AA}\, \dot{n}_{AA,\text{out}}] - \Delta \hat{h}_{r1}\, \dot{n}_{A,\text{in}}. \tag{102}$$

Under most conditions of life, the three terms within brackets are very small in comparison with the last term. Hence, the energy intake can often be approximated as $\dot{E}_i = -\Sigma\, \Delta \hat{h}_{rj}\, \dot{n}_{j,\text{in}} = \Sigma\, \Delta \hat{h}_{j,c}\, \dot{n}_{j,\text{in}}$, where the summation is over all the substrates j entering into combustion reactions j, and $\Delta \hat{h}_{j,c} = -\Delta \hat{h}_{r,j}$ denotes the combustion enthalpy of the substrate. This derivation explains and justifies the common use of combustion enthalpies in the context of describing the energy intake contained in different foodstuffs.

The comprehensive treatment given in (84)–(102) of the simplified situation described in Figure 4.16 shows how complex a detailed bioenergetic analysis must be for real systems. The present treatment calls for a few additional comments.

In the first place, we note that the requirements for the validity of (61) can also be expressed by the statement (cp. Section 4.5.3) that the processes must be isothermal and isobaric and that the summation on the left-hand side of (61) is restricted to contributions from such reactions that are either non-steady or where the components are supplied to or removed from the system. This statement can be verified as follows. If processes 4 and 5 are assumed to be steady, (95) and (97) imply that contributions of the form

$$(\xi_i - \xi_i') \, \Delta \hat{h}_{ri} \tag{103}$$

become zero for $i = 4,5$ so that these reactions do not contribute to (84) for (CV). In the same way, (93) shows that, for $\dot{n}_{FF,out} = 0$ and stationarity of process 3, contribution (103) becomes zero for $i = 3$ in (84). Possible fluxes of F to and from (CV) must then, according to (92), be identical. Finally, (89) shows that, if (103) with $i = 2$ is equal to zero, so that reactions 2 and 2′ proceed at the same rate, AA will deaccumulate, determined by $\dot{n}_{AA,out}$, but this does not affect the total energy balance.

In the second place, we note that the purpose of the analysis has been to determine the total energy expenditure of the system and to determine its distribution in the different parts of the system (subsystems). The practical use of such analyses meets with difficulties. Obviously, according to (86), $\dot{E}$ can be determined by direct calorimetry, but prediction of its value from the left-hand side of (84) requires knowledge of all of the reaction rates and their reaction enthalpies in Table 4.2. The rate ξ_1 can be determined by indirect calorimetry and observation of one of the fluxes in (88), but this is all we can do. The assumption of stationarity for AA, the use of (89) and the observation of $\dot{n}_{AA,out}$ gives the net reaction rate for reaction 2. A similar assumption can be made for reaction 3, cp. (93). The 'hidden' processes 4 and 5 can only be analysed by independent means. Finally, for all internal processes, the heat powers $\dot{Q}_i$ are unknown. The net removal of heat $\Sigma(-\dot{Q}_i)$ can be observed, but its partition among the individual contributions remains unknown.

A detailed bioenergetic analysis of the internal processes in living systems thus requires additional information in terms of biochemical models, models of efficiency, and local experimental observations.

5

Increase of entropy (the second law)

As are the conservation of mass and of energy, so the increase of entropy (the second law of thermodynamics) is a general law, which is independent of the constitution of the medium in question.

The first law deals with energy exchange and energy transformations in processes and ensures – irrespective of the course of the processes – conservation of energy. The second law is not a conservation law, but a law that expresses whether or not a given process will proceed spontaneously. In addition, the second law provides a measure of the deviation of the process from that of ideal behaviour. This deviation is expressed in terms of the entropy production or the irreversibility of the process. In this way, we obtain, together with the first law, (i) a definition of the concept of equilibrium (Chapter 6); (ii) a general expression for the driving forces of processes (Chapter 7); and (iii) a measure of the capacity of processes to exchange useful energy through mutual coupling (Chapter 8).

The second law involves the state property entropy S, which was introduced through the relation $dS = \delta Q_{rev}/T$ for a reversible process in a control mass. In real (irreversible) processes, a larger increase in entropy is observed, as expressed in (2.6) for a homogeneous closed system (CM):

$$dS \geq \delta Q/T. \tag{2.6}$$

This general law is the point of departure of what follows.

The physical meaning of entropy is expressed more clearly by statistical mechanics (see, e.g., Giedt, 1971). Here, entropy is defined as being proportional to the logarithm of the number of microstates by which the particles of the system can attain a given macrostate. There is only one microstate (perfect order) at absolute zero, and $S = 0$ for $T = 0$, as expressed in the third law of thermodynamics ((2.7)). At increasing temperature, the entropy increases and, consequently, it can be perceived as a measure of the degree of disorder.

The fact that the entropy increases by reversible supply of energy as heat, but not as work, is explained by the differences in these forms for energy in transit, as described earlier (in Section 3.6 following Example 3.8). Among these differences, we stress here those associated with energy supplied in the form of electromagnetic radiation δQ_r. They arise because of differences in the system that receives the radiation. In the case of 'unstructured' absorption of thermal radiation, the supply δQ_S is included in δQ in (2.6), just as is the supply of energy by heat conduction or convection. On the other hand, in the case of selective absorption of radiation in a system, with transformation into electrical potential energy (photoelectric effect) or transformation into internal energy (photosynthesis), the energy of the radiation δQ_u should not be included in δQ (to be illustrated in Section 5.4.3). These considerations have no implications for the first law, in which all the energy supplied must simply be accounted for in order to ensure energy conservation.

Using the basal form (2.6) for a closed system (CM), we first derive the more general and useful form for an open system (CV). Instead of using these inequalities, we introduce more convenient equations expressing equalities, denoted entropy balances, which appear after adding terms for entropy production. The subsequent applications include thermo-mechanical, chemo-thermo-mechanical and electro-chemical processes, concluding with the model of a living system described in Figure 4.6.

Note that in classical thermodynamics, entropy production and irreversibility are phenomenological quantities which, in some cases, can be determined indirectly from the entropy balance. Direct determination of these quantities can be obtained by the use of the theory of irreversible thermodynamics (Chapter 7), provided that fluxes and driving forces in the constitutive relations are known for the processes in question.

5.1 The second law for control mass and control volume

The increase in entropy for an unsteady process in a control mass is obtained from (2.6), after division by dt and carrying out the usual limit process $\Delta t \to 0$, yielding

$$d(Ms)/dt \geq \dot{Q}/T. \tag{1}$$

The entropy of the control mass thus increases at least at the rate given by the supplied heat power $\dot{Q}$ divided by the absolute temperature of the control mass at the given time. When heat is removed, $\dot{Q}$ is negative and the entropy of the system may decrease, but not faster than is given by (1). For a *reversible process*, the equality sign in (1) is valid. In all cases, according to the concept of state, (1) assumes that, at any time, the system is in a homogeneous equilibrium state so that the entropy function S and the absolute temperature T exist as state properties.

For a process $1 \rightarrow 2$, integration of (1) over time, using mass conservation $M_1 = M_2 = M$ from (3.8), gives

$$M(s_2 - s_1) > \int_1^2 (1/T) \, \delta Q \quad \text{(CM)}. \tag{2}$$

For an isothermal process, $T = $ constant and the right-hand side of (2) is $_1Q_2/T$. For an adiabatic process, $\delta Q = 0$ and the right-hand side is zero. In other cases, calculation of the right-hand side of (2) requires knowledge of $\delta Q(T)$, or $\dot{Q}(t)$ and $T(t)$, over the process. Form (2) is valid for an arbitrary process, regardless of whether or not it proceeds through equilibrium states, provided only that the initial and final states are equilibrium states.

For a *cycle*, (2) reduces to

$$0 > \oint \delta Q/T \quad \text{(CM)}, \tag{3}$$

since this process, by definition, ends at the initial state.

The increase of entropy for a control volume is obtained by transformation of the left-hand side of (1) using the Reynolds transport theorem ((3.5)):

$$d(Ms)dt + \Sigma \, (\dot{m}s)_{\text{out}} - \Sigma \, (\dot{m}s)_{\text{in}} \geq \dot{Q}/T \quad \text{(CV)}. \tag{4}$$

The increase over time of entropy of the control volume plus the net outflow of entropy is therefore not less than the supply of heat divided by the absolute temperature of the control volume. The equality sign in (4) is valid for a reversible process. For a steady process, the first term in (4) is zero.

We may employ the general form (4) for any system since it includes and reduces to (1) for the control mass when no mass crosses the control surface. Use of molar state properties in (4) gives

$$d(N\hat{s})/dt + \Sigma_{\text{out}}\dot{n}\hat{s} \geq \dot{Q}/T, \tag{5}$$

where $N = M/\hat{M}$ and $\dot{n} = \dot{m}/\hat{M}$, and the sum 'out' implies 'net out', cp. (4.9).

5.2 The entropy balance. Entropy production and irreversibility

The form of the second law as an inequality suggests that a non-negative source term $\dot{\sigma}$ is missing for this general law to have the usual form of a balance equation ((3.6)). The source term $\dot{\sigma}$ defines the *entropy production* of the system. This is a quantitative measure of the magnitude of the irreversibilities that are associated with the processes taking place *within* the surface of the control volume. For irreversible processes $\dot{\sigma}$ is positive and is zero for reversible processes, i.e. $\dot{\sigma} \geq 0$.

The entropy balance can be perceived as a *derived general law*, which follows from, and is an alternative form of, the second law.

Addition of $\dot{\sigma}$ to the right-hand side of (4) gives the entropy balance for a control volume:

$$d(Ms)/dt + \Sigma\,(\dot{m}s)_{\text{out}} - \Sigma\,(\dot{m})_{\text{in}} = \dot{Q}/T + \dot{\sigma} \quad \text{(CV)}. \tag{6}$$

For a differential process in a closed system (CM), multiplication of (6) by dt gives

$$dS = \delta Q/T + \delta\sigma \quad \text{(CM)}, \tag{7}$$

which is analogous to the addition of the differential entropy production $\delta\sigma$ to (2.6) in order to produce a balance equation. When (7) is multiplied by the absolute temperature T, we obtain

$$T\,dS = \delta Q + \delta I; \quad \delta I \equiv T\,\delta\sigma, \tag{8}$$

where δI (≥ 0), having the unit of energy, is the differential *irreversibility* of the process. In a similar manner,

$$\dot{I} = T\,\dot{\sigma} \tag{9}$$

represents the irreversibility power (energy per unit time) for an arbitrary steady or unsteady process. The term appears after multiplication of (6) by the absolute temperature. This idea is pursued further when a general balance equation for the available energy (exergy) is derived (Section 8.2).

On the basis of the second law and the definitions of entropy production and irreversibility, we can define some useful additional concepts. Firstly, a *reversible process* can be characterized by $\delta\sigma = 0$, $\delta I = 0$ and $\dot{I} = 0$. Secondly, a *reversible, adiabatic process* is also an *isentropic process*, i.e. a process during which the entropy remains constant. This is immediately clear from (8), where $\delta I = 0$ and $\delta Q = 0$

implies that $dS = 0$, or $S =$ constant for a reversible, adiabatic process in a control mass. In an analogous manner, (6), with $\dot{\sigma} = 0$ and $\dot{Q} = 0$, implies that *entropy is conserved* in a reversible, adiabatic process in a control volume. For a stationary process this means that the entropy flows out and in are identical; in the case of one flow in and one flow out (e.g. through a pump), the specific entropy is constant from inlet to exit.

5.3   Thermal and mechanical processes

In this section we deal with a number of thermal and mechanical processes for media that can be considered as consisting of one substance.

As an introduction, we derive a few general results with respect to a differential process in a control mass in which there are no changes in kinetic and potential energy.

This condition is no real restriction on the derived results. It can be shown that an increase in kinetic and potential energy can always be associated with a given supply of mechanical work δW_m by way of a mechanical energy balance. If such contributions are subtracted from the first law, we obtain the balance of internal energy. Here, as shown in the following, δW denotes the remaining energy supply as work. Recall also the remarks following (3.36).

From the first law, the Gibbs relation ((3.36)) and the entropy balance on form (8), i.e.

$$dU = \delta Q + \delta W; \quad dU = T\,dS + \delta W_{rev}; \quad T\,dS = \delta Q + \delta I, \quad (10)$$

we obtain, after elimination of δQ and TdS,

$$\delta I = \delta W - \delta W_{rev} = \delta W_{irr}. \quad (11)$$

In this case, the differential irreversibility can therefore be associated with *irreversible work*, i.e. either some real work associated with friction (Figure 3.8(*b*)), or some fictitious work associated with the process (Figure 3.7(*a*)).

Processes in incompressible, thermomechanical media are of considerable practical interest. Here, for a control mass, (3.38) and (3.60) reduce to

$$dU = T\,dS; \quad dU = Mc\,dT, \quad (12)$$

which, using (8), gives

$$dU = Mc\,dT = \delta Q + \delta I, \quad (13)$$

which is interpreted as follows. Firstly, for an isothermal process, $dT = 0$ in (13), irreversibilities within the system are transformed into energy removed as heat:

$$\delta I = - \delta Q \text{ (incompressible, isothermal).} \tag{14}$$

The first law from (10), with $dU = 0$ for an isothermal process, gives $\delta W = - \delta Q$, which is in agreement with (11), since the reversible work for an incompressible, thermomechanical medium is equal to zero. Secondly, for an adiabatic process, $\delta Q = 0$ in (13), the irreversibilities are transformed into an increase in internal energy:

$$\delta I = dU \text{ (incompressible, adiabatic).} \tag{15}$$

Note that (11) includes no contribution to irreversibility associated with transfer of energy as heat. This is so because our analysis assumes uniform states, including a uniform temperature, within the control mass.

Figure 5.1 shows a composite isolated system with heat transfer across a finite temperature difference, $\Delta T = T_1 - T_2 > 0$. Here, δQ is transferred from $(CM)_1$ at T_1 to $(CM)_2$ at T_2. Both control masses have a uniform temperature and there are no contributions from work energies, so that the first and second laws give

$$dU_1 = - \delta Q; \quad dS_1 = - \delta Q/T_1 \text{ (CM)}_1, \tag{16}$$

$$dU_2 = \delta Q; \quad dS_2 = \delta Q/T_2 \text{ (CM)}_2. \tag{17}$$

Thus, there is no entropy production within either of the two control masses. On the other hand, if we were to consider the composite system (CM) in Figure 5.1, we would obtain

$$d(U_1 + U_2) = 0; \quad d(S_1 + S_2) = \delta\sigma, \tag{18}$$

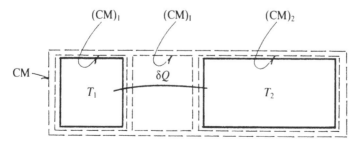

Fig 5.1 A composite isolated system (CM), consisting of $(CM)_1$ at T_1 and $(CM)_2$ at T_2, which exchange energy as heat through a medium $(CM)_I$ without mass.

which, using (16) and (17), reveals the differential entropy production

$$\delta\sigma = \delta Q(1/T_2 - 1/T_1) \geqslant 0, \tag{19}$$

or, after division by the time interval dt, the instantaneous rate

$$\dot{\sigma} = \dot{Q}(1/T_2 - 1/T_1) \geqslant 0. \tag{20}$$

This result is also obtained by considering $(CM)_I$, which may be a region without mass between $(CM)_1$ and $(CM)_2$ and which localizes the entropy production. Result (19) or (20) shows that $T_1 > T_2$ for $\delta Q > 0$. This is in agreement with our experience and it shows how the second law dictates the possible process or direction of a given process.

Figure 5.1 can also be used to illustrate the concept of a *given system*, e.g. $(CM)_2$, the *surroundings*, for e.g. $(CM)_1$, and the *universe* (CM), i.e. everything of importance, since (CM) is isolated. For $T_1 > T_2$ we saw that the entropy of the given system $(CM)_2$ increases, while it decreases for the surroundings $(CM)_1$, but increases in the universe. For the isolated system (the universe), which must be adiabatic, (7) also shows that the total entropy must increase, $dS > 0$.

Example 5.1

The convective heat flux $\dot{Q}/A$ across the finite temperature drop $T_w - T_b$ in Example 3.11 gives, for the surface area $A = 1\,\mathrm{m}^2$, the total entropy production

$$\dot{\sigma} = \dot{Q}(1/T_b - 1/T_w) = 50(1/273 - 1/277.5) = 0.42\,\mathrm{mW/K}. \tag{a}$$

When heat is supplied to, or removed from, a solid body of finite dimension, the process is accomplished by convection. Temperature gradients must appear within the body so that the heat flux supplied at the boundary can be distributed throughout the system. In most biological systems, the associated entropy production is small, because the temperature gradients are small, and it can often be neglected. As shown in the following example, the local entropy production per unit volume by one-dimensional heat conduction in the x-direction can be calculated as

$$\dot{\sigma}/\mathcal{V} = (\dot{Q}/A)\,d(1/T)/dx = (k/T^2)\,(dT/dx)^2, \tag{21}$$

where the heat flux $\dot{Q}/A$ is given by (3.26), and k denotes the heat conductivity of the medium.

Example 5.2

We wish to calculate the total entropy production and the irreversibility associated with the one-dimensional, steady heat conduction through the blubber of the seal in Example 3.10.

Figure 5.2 shows the differential $(CM)_d$ of thickness dx. During a differential time interval dt, a quantity of heat δQ is removed and supplied since $dU = 0$ for stationarity. For the same reason, $dS = 0$ and (7) gives

$$0 = \delta Q/T - \delta Q/(T + \delta T) + \delta \sigma, \tag{a}$$

or, for $\delta T/T \ll 1$,

$$\delta \sigma = \delta Q(-\delta T/T^2). \tag{b}$$

Division by dt and the volume $\delta \mathcal{V} = A\, dx$ gives

$$\dot{\sigma}/\mathcal{V} = (\dot{Q}A)_x\, d(1/T)/dx = (k/T^2)\,(dT/dx)^2, \tag{c}$$

where $(\dot{Q}/A)_x = -k\, dT/dx$ from (3.26) and $\delta T = (dT/dx)dx$. Result (c) is (21) and can be seen to be always positive, in agreement with the second law. This correct result is, of course, due to the fact that Fourier's law ((3.26)) has been formulated correctly, in agreement

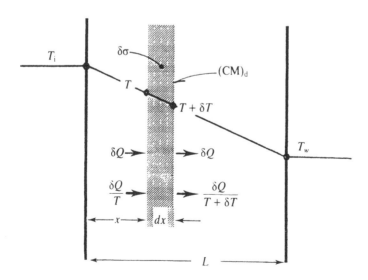

Fig 5.2 Steady heat conduction through wall with differential control mass $(CM)_d$.

with the second law. That is, the heat flux in the x-direction is positive for a negative gradient, i.e. a decreasing temperature in that direction. The entropy production per unit area of the wall in Example 3.10 is now obtained by integration of (c) over the thickness L with

$$T(x) = T_i + (\mathrm{d}T/\mathrm{d}x)(x - 0); \quad \mathrm{d}T/\mathrm{d}x = -(T_i - T_w)/L = \text{constant}, \tag{d}$$

which gives, per unit area

$$\dot{\sigma}/A = k(T_i - T_w)(1/T_w - 1/T_i), \tag{e}$$

or, multiplying by A and using (3.26), the total entropy production

$$\dot{\sigma} = \dot{Q}(1/T_w - 1/T_i) = 50 \times (1/277.5 - 1/310) = 0.019 \text{ W/K}. \tag{f}$$

In this special case, we obtain the same expression as (a) in Example 5.1

The total irreversibility is obtained by use of (9), (c) and (d):

$$\dot{I} = A \int_0^L T \,(\sigma/\mathscr{V}) \, \mathrm{d}x = A[k(T_i - T_w)/L] \ln (T_i/T_w), \tag{g}$$

$$\dot{I} = 1 \times 0.1 \times [(37 - 4.5)/0.065] \ln (310/277.5) = 5.54 \text{ W}, \tag{h}$$

which is later (in Section 8.3) shown to be the loss of useful energy, here about 11% of the transmitted energy of 50 W.

Example 5.3

As shown by the use of (7) in (a) in Example 5.2, the term $\delta Q/T$ is the sum of contributions Q/T, also called the *entropy flux*, supplied to the system and evaluated at the local system temperature along its boundary. For a system with non-uniform temperature, the term should actually be written as $\delta(Q/T)$, as noted by (2.6), meaning a net supply of entropy flux. The corresponding entropy balance is

$$\mathrm{d}S = \delta(Q/T) + \delta\sigma, \tag{a}$$

or, after division by $\mathrm{d}t$, the rate

$$\mathrm{d}S/\mathrm{d}t = \Sigma_{\text{in}} \,(\dot{Q}/T) + \dot{\sigma}. \tag{b}$$

By the rule of differentiation, the term $\delta(Q/T)$ is expresssed as the sum of two terms:

$$\delta(Q/T) \quad = \quad \delta Q/T \quad + \quad Q\delta(1/T). \qquad (c)$$

| net supplied entropy flux | reversible increase of entropy at the mean system temperature | irreversible entropy production due to the transfer of heat over a temperature drop |

The above interpretation of (c) is verified by writing, for the actual incompressible medium, the Gibbs relation and the first law:

$$dU = TdS; \quad dU = \delta Q, \qquad (d)$$

from which we derive the reversible contribution $\delta Q/T = dS$ (since T in the Gibbs relation is the uniform (average) temperature). Then (a) gives

$$\delta\sigma = -Q\delta(1/T), \qquad (e)$$

or, in rate form per unit volume,

$$\dot{\sigma}/\mathcal{V} = (\dot{Q}/A)_x \, d(1/T)/dx, \qquad (f)$$

which is (c) in Example 5.2. The negative sign in (e), apparently in conflict with (b) in Example 5.2, is to be understood as follows. According to convention, δ in (a), (c), (d), and hence also in (e), implies a net supply, i.e. a net flux in, while d/dx signifies a net flux out of a differential system.

Example 5.4

The entropy production during the instantaneous and adiabatic expansion of the ideal gas in Example 4.6 to twice its volume can be calculated by integration of (7), and use of (b) in Example 4.6:

$$_1\sigma_2 = M(s_2 - s_1) = MR \ln(\mathcal{V}_2/\mathcal{V}_1), \qquad (a)$$

which is positive for the spontaneous process where $\mathcal{V}_2/\mathcal{V}_1 > 0$. Since $u = $ constant implies that $T = $ constant, integration of (8) gives the irreversibility

$$_1I_2 = T \, _1\sigma_2. \qquad (b)$$

For the reversible, adiabatic expansion with performance of work (Figure 3.7(b)) (7) gives $S = $ constant, as expected for an isentropic process.

The well-known isentropic relations for an ideal gas with constant heat capacities, $c_p/c_v = \gamma$:

$$T/T_1 = (p/p_1)^{(\gamma-1)/\gamma} = (v/v_1)^{-(\gamma-1)} \tag{c}$$

were derived in (g) and (h) of Example 4.6.

Example 5.5

Heating of an ideal gas in a rigid container (Figure 3.8(a)) was analysed by the first law, yielding (a) in Example 4.7. Integration of (7) and (8) gives the additional relations

$$M(s_2 - s_1) = \int_1^2 \delta Q/T + {}_1\sigma_2; \quad M \int_1^2 T \, ds = {}_1Q_2 + {}_1I_2. \tag{a}$$

When the process advances through equilibrium states, so that the temperature of the gas is uniform at all times, it must be reversible, ${}_1\sigma_2 = 0$ and ${}_1I_2 = 0$, and use of (3.44) with $dv = 0$ (rigid container) in the second equation in (a) gives

$${}_1Q_2 = Mc_v(T_2 - T_1), \tag{b}$$

which is (a) in Example 4.7.

When the same energy is supplied as work (Figure 3.8(b)), (a) gives

$${}_1I_2 = Mc_v(T_2 - T_1),$$

which, using (b) in Example 4.7, shows that the irreversibility is equal to the energy supplied as work, which is 'lost' in the process.

Example 5.6

For heating of the steady flow of air in Example 4.8, the entropy balance for CV in Figure 4.6 (without the addition of water vapour) is given by (6):

$$\dot{m}(s_2 - s_1) = \int \delta \dot{Q}/T + \dot{\sigma}. \tag{a}$$

The left-hand side of (a) can be calculated directly from (3.58) to $\dot{m}c_p \ln (T_2/T_1)$, when we neglect the pressure drop due to friction (and thereby its contribution to entropy production). On the other hand, the integral over CV in (a) can only be calculated after specification of the temperatures associated with the heat transfer.

In this example, we can assume that heat is supplied from tissues at $T_t = 37\,°C = 310\,K$ during the whole process. In this case we obtain from (a)

$$\dot{\sigma} = \dot{m}c_p \ln (T_2/T_1) - \dot{Q}/T_t. \tag{b}$$

Inserting values from Example 4.8 gives

$$\begin{aligned}\dot{\sigma} &= 1.29 \times (0.006/60) \times 1000 \ln (310/273) - 4.8/310\\ &= 0.0164 - 0.0155 = 0.91\ mW/K, \end{aligned} \tag{c}$$

or, with respect to T_t, the irreversibility

$$\dot{I} = T_t\,\dot{\sigma} = 5.08 - 4.8 = 0.28\ W, \tag{d}$$

which is about 6% of the transmitted heat power of the 4.8 W. It should be noted that a choice of $T < T_t$ in (b) would give a lower entropy production.

5.4 Chemical–thermal–mechanical processes

For these important processes in living systems we first generalize the second law ((5)), or the more convenient entropy balance ((6)), in order to describe one or several steady or non-steady chemical reactions. On this basis we then calculate the entropy production and the irreversibility, quantities that must always be positive for real processes. Finally, we deal with coupled processes and photosynthesis.

5.4.1 Generalization

With constant (T,p) and use of constant average values for the partial molar entropies $\bar{s}$, we obtain (6) on a molar basis (Figure 5.3):

$$\Sigma \hat{\bar{s}}_j\, dN_j/dt + \Sigma_{out} \Sigma\,\dot{n}_j\,\hat{\bar{s}}_j = \dot{Q}/T + \dot{\sigma}, \tag{22}$$

$$dN_j/dt + \Sigma_{out}\,\dot{n}_j = \nu_j\xi. \tag{23}$$

Here, the mole balance ((4.58)) is given in (23) for a single chemical reaction with reaction rate ξ and stoichiometric coefficients ν_j defined by (3.77).

Equation (22) now contains absolute values for the state properties of the mixture (cp. Section 4.5.1), expressed on a molar basis by summation of the contributions of the individual components. Even for

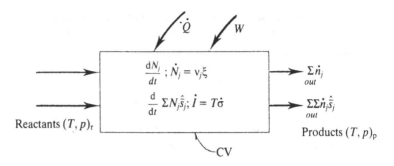

Fig 5.3 Unsteady control volume process with a single chemical reaction; mole balance for component j and entropy balance for the total mixture.

ideal mixtures, the partial molar entropy $\hat{\bar{s}}$ for a component depends on the actual composition of the mixture, cp. (3.101). It is therefore necessary to know if the individual components are supplied, or removed, as pure substances, or if they are parts of mixtures (gas, liquid or solid), and, if so, the composition of these mixtures. Furthermore, changes over time of the composition of the mixture in the control volume will change the values of the molar entropies, also at constant (T,p). In analyses of processes in biological systems it is customary to use constant average values for $\hat{\bar{s}}$ in homogeneous mixtures, or $\hat{s}$ for a pure substance in heterogeneous mixtures, since the errors involved will usually be small. With constant (T,p), we can therefore assume that $d\hat{\bar{s}}_j/dt = 0$, and this is used in the first term in (22).

We now follow the derivation in Section 4.5.3 by multiplying (23) by $\hat{\bar{s}}$ summing over all components and subtracting (22) from the result. In analogy with (4.60), we obtain

$$\dot{\xi}\,\Delta\hat{s}_r = \dot{Q}/T + \dot{\sigma}, \qquad (24)$$

where the *molar reaction entropy* is defined by

$$\Delta\hat{s}_r = \Sigma\,\nu_j\hat{s}_j, \qquad (25)$$

with $\nu_j = -1$ for $j =$ substrate and the components as pure substances. It should be noted that definition (25) implies that the reaction entropy is a state property, dependent on (T,p) only, and that contributions from entropy of mixing are consequently neglected in (24). For combustion of glucose ((4.46)), (25) takes the form

$$\Delta\hat{s}_r = 6\,\hat{s}(CO_2) + 6\,\hat{s}(H_2O) - \hat{s}(G) - 6\,\hat{s}(O_2). \qquad (26)$$

When several reactions proceed simultaneously with reaction rates $\dot{\xi}_i$

and reaction entropies $\Delta \hat{s}_{ri}$, $i = 1,2,\ldots,r$, it follows, using (3.80), that (24) can be generalized to

$$\Sigma_r \, \xi_i \, \Delta \hat{s}_{ri} = \dot{Q}/T + \dot{\sigma}. \tag{27}$$

As for (4.61), (27) is valid in the case of isothermal and isobaric reactions, and also when there is accumulation (deaccumulation) of the components in the control volume. However, there must be stationarity with respect to possible intermediaries X_i and components Y_i in coupled reactions within the control volume (cp. Figure 4.12). The summation on the left-hand side of (27) must thus include contributions from all reactions that are either non-stationary or for which one or several reactants/products is supplied to, or removed from, the control volume.

As illustrated in the preceding section, the second law can be used to calculate the entropy production $\dot{\sigma}$ or the irreversibility $\dot{I}$, which are both non-negative. Use of (27) requires knowledge of $\dot{Q}$, which, however, is often unknown. It is therefore of interest that $\dot{Q}$ can be eliminated by use of the first law (4.61). By this step, the work power $\dot{W}$ now appears, but this is often known, and (27) then gives an alternative expression for the entropy production:

$$\dot{\sigma} = (1/T) \, [\dot{W} - \Sigma_r \, \xi_i \, (\Delta \hat{h}_{ri} - T \, \Delta \hat{s}_{ri})], \tag{28}$$

or, after multiplication by the absolute temperature, the irreversibility rate

$$\dot{I} = \dot{W} - \Sigma_r \, \xi_i (\Delta \hat{h}_{ri} - T \, \Delta \hat{s}_{ri}). \tag{29}$$

The expressions within parentheses in (28) and (29) represent the change in the Gibbs function of the ith reaction at temperature T, i.e. the *reaction Gibbs function*:

$$\Delta \hat{g}_{ri} = \Sigma \, \nu_{ij} \, \hat{\mu}_j = \Delta \hat{h}_{ri} - T \, \Delta \hat{s}_{ri}. \tag{30}$$

Hence, (29) can also be written

$$\dot{I} = \dot{W} - \Sigma_r \, \xi_i \, \Delta \hat{g}_{ri}. \tag{31}$$

As pointed out earlier, cp. (3.86), the Gibbs function is merely a derived thermodynamic state property, which, as we now see, it is convenient to use in the treatment of chemically reactive mixtures. Appendix C.1 gives values for enthalpy of formation, the Gibbs function of formation, absolute enthalpy and heat capacity for selected substances. Appendix C.2 gives values of standard reaction enthalpy and standard reaction

Gibbs function for selected reactions. For definitions of these quantities and their conversion to other states, we refer to Sections 4.5.1 and 3.10.5.

The preceding derivation is analogous to that given by (10)–(11) for thermomechanical processes in a control mass, and comparison of (31) with (11) suggests that the last term in (31) represents the possible reversible work power for the process. This expectation is confirmed by considering a reversible process, $\dot{I} = 0$, for which (31) is reduced to

$$\dot{W}_{rev} = \Sigma_r \, \xi_i \, \Delta \hat{g}_{ri}. \tag{32}$$

In addition to these general remarks, we note that for some of the most important reactions in biological systems, i.e. the combustion reactions mentioned in Section 4.5.4, the last term in (30) is very small in relation to the first term so that

$$\Delta \hat{g}_r \approx \Delta \hat{h}_r. \tag{33}$$

In these cases it follows, from (27) and (4.61), that the removed heat power is approximately equal to the irreversibility rate of a system:

$$-\dot{Q} \approx \dot{I}. \tag{34}$$

This result is true for most animal species. Disregarding growth, and considering, therefore, stationary states of the whole organism, the performance of external work $(-\dot{W})$ is generally much smaller than the heat losses $(-\dot{Q})$. This means that only a very small part of the useful chemical energy is used for external work. The overwhelming part is used to maintain processes within the body and is dissipated therein before leaving the body as heat. It should be noted that (34) is an expression of the sum of all processes which have been assumed to be isothermal, and it has nothing to do with such irreversibilities that, as calculated in (21), are associated with heat conduction within the body. Temperature differences within the body of homeothermic animals are small, generally less than 5°C, and the associated irreversibilities are very small.

As noted in the introduction, the second law determines whether or not a given process proceeds spontaneously. This is the case for a single process whenever the entropy production (or the irreversibility) is positive. However, the *rate* of the process is determined by other factors, e.g. the presence of catalysts, and will not be treated in this text.

Example 5.7

Continuing from Example 4.15, we wish to calculate the entropy of reaction for the isobaric and isothermal combustion of glucose, and the entropy production and irreversibility at a consumption of 389 g/day. The reaction is assumed to take place in tissues (i.e. in aqueous solution) at 1.0 atm and 37 °C, and the steady composition of the mixture is, as in Example 4.15, given by 0.01 mol/l glucose and partial pressures of carbon dioxide and oxygen of 0.07 and 0.21 atm respectively. The state of the mixture is thus characterized by the activity ratios (cp. Section 3.10.5) a_j/a_j^0 = 0.07, 1.0, 0.01 and 0.21 for carbon dioxide, water, glucose and oxygen, respectively.

Using the reaction equation ((4.46)) and the modification that all components are in aqueous solution, (26) is used with values from Appendix C.1 at (T_0,p_0) for the components as pure substances in the standard state:

$$\Delta \hat{s}_r^0 = 6 \times 121 + 6 \times 70 - 151 - 6 \times 115 = 305 \text{ J/mol-K.} \qquad \text{(a)}$$

This reaction entropy must be converted into that of the actual state. Conversion of the entropy for each component at constant temperature to another concentration takes place, as mentioned in Section 3.10.5, by (3.96) and (3.121). When the entropy of mixing is zero, or vanishingly small, we have

$$\hat{s}_j(T,p_0,\chi_j) = \hat{s}_j(T,p_0) - \hat{R} \ln (a_j/a_j^0). \qquad \text{(b)}$$

Conversion of the entropy of the pure components from one temperature to another takes place by (3.58) and (3.61) for an ideal gas or an incompressible liquid or solid:

$$\hat{s}_j(T,p_0) = \hat{s}_j(T_0,p_0) + \hat{c}_p \ln (T/T_0). \qquad \text{(c)}$$

Substitution of (b) and (c) in (25) gives

$$\Delta \hat{s}_r = \Delta \hat{s}_r^0 + \ln (T/T_0) \, \Sigma \, \nu_j \, \hat{c}_p - \hat{R} \, \Sigma \, \nu_j \, \ln (a_j/a_j^0). \qquad \text{(d)}$$

Using numerical values given above and from Example 4.15, we obtain at 25 °C

$$\Delta \hat{s}_r = 305 + 0 - 8.314 \times \ln [(0.07^6 \times 1.0^6)/(0.01 \times 0.21^6)]$$
$$= 305 + 0 - 16.5 = 288.5 \text{ J/mol-K,} \qquad \text{(e)}$$

and, at 37 °C,

$$\Delta \hat{s}_r = 305 + \ln (310/298) \times 279 - 16.5 = 305 + 10.9 - 16.5$$
$$= 299.4 \text{ J/mol-K.} \qquad \text{(f)}$$

Substitution of (a) in place of (e) or (f) thus introduces errors of less than 6%, which is a typical value.

Next, in order to determine the entropy production from (24), we need to know $\dot{Q}$ and T. If we assume that the work output is zero, the first law ((4.60)) shows that the whole energy expenditure is removed as heat, $\dot{E} = (-\dot{Q}) = 71.7\,W$, in accordance with (h) in Example 4.15. Substitution of (g) in Example 4.15, as well as (f) above, in (24) gives

$$\dot{\sigma} = \xi\Delta\hat{s}_r - \dot{Q}/T = 0.025 \times 0.2994 + 71.7/310 = 0.239\,\text{W/K}, \quad (g)$$

and the irreversibility becomes

$$\dot{I} = T\,\dot{\sigma} = 310 \times 0.239 = 74.0\,\text{W}, \tag{h}$$

of which 71.7 W originates from the reaction enthalpy and 2.3 W from the reaction entropy.

Example 5.8

We wish to calculate the reaction Gibbs function for combustion of glucose at 25 °C and at 37 °C.

Using $\Delta\hat{h}_r$ from (b) and (c) in Example 4.15 and $\Delta\hat{s}_r$ from (e) and (f) in Example 5.7 inserted into (30), we obtain, at $T_0 = 25\,°C = 298\,K$:

$$\Delta\hat{g}_r = -2870 - 298 \times 0 \times 289 = 2870 - 86.0 = 2956\,\text{kJ/mol}, \quad (a)$$

and, at 37 °C = 310 K:

$$\Delta\hat{g}_r = -2867 - 310 \times 0 \times 299 = -2867 - 92.8 = -2960\,\text{kJ/mol}. \tag{b}$$

In both cases, use of (33) would cause an error of about 3%.

Example 5.9

We wish to show that combustion of hydrogen (g) with oxygen (g) to produce water (aq) at 25 °C is an irreversible process.

The reaction entropy for the isobaric and isothermal process in Figure 4.11 is, according to (25) and values from Appendix C.1 at 25 °C,

$$\Delta\hat{s}_r^0 = \hat{s}^0(H_2O) - \hat{s}^0(H_2) - \tfrac{1}{2}\,\hat{s}^0(O_2) = 70 - 131 - (\tfrac{1}{2}) \times 205$$
$$= -163.5\,\text{J/mol-K}. \tag{a}$$

The reduction in entropy for this process is due to the fact that heat is removed, as calculated in (b) in Example 4.14. Substitution of this result, and (a) above, in (24) shows that the process is irreversible, since the molar entropy production is positive:

$$\dot{\sigma}/\dot{n}(H_2O) = -0.1635 - (-286/298) = 0.796 \text{ kJ/K-mol } H_2O \quad \text{(b)}$$

as is the irreversibility,

$$\dot{i}/\dot{n}(H_2O) = T \dot{\sigma}/\dot{n}(H_2O) = 237.3 \text{ kJ/mol } H_2O. \quad \text{(c)}$$

For this example, the first and the second laws show that the irreversibility is the negative of the reaction Gibbs function:

$$\Delta\hat{g}_r^0 = \Delta\hat{h}_r^0 - T \Delta\hat{s}_r^0 = (-286) - 298 \times (-0.1635)$$
$$= -237.3 \text{ kJ/mol } H_2O. \quad \text{(d)}$$

The example also shows that there is a large difference (about 20%) between $\Delta\hat{h}_r$ and $\Delta\hat{g}_r$, so that approximation (33) is unacceptable.

Example 5.10

The reaction entropy and the reaction Gibbs energy are sought for the reaction

$$ATP \rightarrow ADP + P, \quad \text{(a)}$$

when this proceeds in a cell (in aqueous solution) at 310 K and at concentrations of ATP, ADP and P of 4, 1 and 10 mmol/l respectively.

At the standard state (each component at a concentration of 1 mole per l, cp. Table 3.3 in Section 3.10.5), Appendix C.1 gives the following values at 310 K:

$$\Delta\hat{h}_r^0 = -20 \text{ kJ/mol-ATP} , \ \Delta\hat{g}_r^0 = -31 \text{ kJ/mol-ATP}, \quad \text{(b)}$$

from which, with (30),

$$\Delta\hat{s}_r^0 = [(-20) - (-31)]/310 = 0.0355 \text{ kJ/K-mol-ATP}. \quad \text{(c)}$$

Conversion of the reaction Gibbs energy to actual concentrations at the same temperature takes place by using (3.121), with (3.125) and $y_i = 1$, for each component, inserted into the summation in (30):

$$\Delta\hat{g}_r = \Delta\hat{g}_{r0} + \Sigma \ \nu_j\hat{R}T \ln (c_j/c_j^0), \quad \text{(d)}$$

which gives, using (b) and $\nu(ATP) = -1$, $\nu(ADP) = 1$ and $\nu(P) = 1$ for reaction (a):

$$\Delta \hat{g}_r = \Delta \hat{g}_r^0 + \hat{R}T \ln [c(ADP) \times c(P)/c(ATP) \times c^0]$$
$$= -31 + 8.314 \times 10^{-3} \times 310 \times \ln [10^{-3} \times 10^{-2}/4 \times 10^{-3} \times 1]$$
$$= -46 \text{ kJ/mol-ATP}. \tag{e}$$

Using this value, and the reaction enthalpy from (b), which with the assumption of ideality is independent of changed concentrations, (30) gives

$$T \Delta \hat{s}_r = -20 + 46 = 26 \text{ kJ/mol-ATP}, \quad \Delta \hat{s}_r = 84 \text{ J/mol-K}. \tag{f}$$

5.4.2 Coupled chemical reactions

The concept of coupled chemical reactions was introduced in Section 4.5.5 to deal with the simultaneous occurrence of two or more reactions where one does not proceed spontaneously but requires the supply of energy from other reactions. This supply is a transfer of internal chemical energy in the form of useful energy (Chapter 8).

For the example of two coupled reactions, combustion of glucose ((4.46)) and synthesis of ATP ((4.69)), treated together (Figure 4.14(a)), the energy balance ((4.68)) and the entropy balance ((27)) give

$$\xi_1 \Delta \hat{h}_{r,1} + \xi_2 \Delta \hat{h}_{r,2} = \dot{Q} \text{ (CV)}, \tag{35}$$

$$\xi_1 \Delta \hat{s}_{r,1} + \xi_2 \Delta \hat{s}_{r,2} = \dot{Q}/T + \dot{\sigma} \text{ (CV)}. \tag{36}$$

In so far as these equations are fulfilled by $\dot{\sigma} > 0$, we conclude that the total process is thermodynamically possible.

When the total process is considered as consisting of two coupled reactions with a chemical energy transfer $\dot{W}_{Ch}$ (Figure 4.14(b)), we obtain for the two control volumes

$$\xi_1 \Delta \hat{h}_{r,1} = \dot{Q}_1 - \dot{W}_{Ch}; \quad \xi_1 \Delta \hat{s}_{r,1} = \dot{Q}_1/T + \dot{\sigma}_1 \text{ (CV)}_1, \tag{37}$$

$$\xi_2 \Delta \hat{h}_{r,2} = \dot{Q}_2 + \dot{W}_{Ch}; \quad \xi_2 \Delta \hat{s}_{r,2} = \dot{Q}_2/T + \dot{\sigma}_2 \text{ (CV)}_2, \tag{38}$$

for which it must be true that $\dot{Q}_1 + \dot{Q}_2 = \dot{Q}$ and $\dot{\sigma}_1 + \dot{\sigma} = \dot{\sigma}$, and, furthermore, that $\dot{\sigma}_1 > 0$ and $\dot{\sigma}_2 > 0$, which implies restrictions on the magnitudes of $\dot{Q}_1$ and $\dot{Q}_2$. Use of (37) and (38) requires knowledge of $\dot{W}_{Ch}$ or the partition of $\dot{Q}$ between $\dot{Q}_1$ and $\dot{Q}_2$ (as illustrated later in Example 8.6).

5.4.3   Photosynthesis

This process occurs, as described in Section 4.5.6, in green plants. Of the total supplied radiation energy $\dot{Q}_r$, part, $\dot{Q}_u$, is directly transformed into internal chemical energy bound in glucose according to reaction (4.72). Hence $\dot{Q}_u$ does not contribute to the right-hand side of the entropy balance ((24)), just as the internal energy transfer in coupled reactions does not contribute.

In a plant, considered as an isothermal control volume, the irreversibility is determined solely by the chemical reactions and the transport processes within it. Outside the plant, between the sun and the plant (in association with the supply of radiation energy) and between the plant and its surroundings (in association with the heat losses), there is a further entropy production which can be calculated as explained in Example 5.1.

Example 5.11

We calculate the irreversibility over a time interval of eight hours for the green leaf in Example 4.19. This leaf receives radiation energy $\dot{Q}_s \approx 500\,\text{mW}$ and gives up heat power $\dot{Q}_1 = 490\,\text{mW}$ in an isothermal process at $T = 25\,°\text{C} = 298\,\text{K}$ with synthesis of glucose according to (4.72).

Taking the leaf as a control volume, the first law gave result (b) in Example 4.19, i.e.

$$\xi \, \Delta\hat{h}_r = \dot{Q}_s - \dot{Q}_1, \tag{a}$$

which shows that the net supply of energy as heat, $\dot{Q}_u = \dot{Q}_s - \dot{Q}_1$, is equal to the part which is transformed directly into internal chemical energy in glucose. This contribution is therefore not accounted for in (24), which thus shows that the entropy production is determined by the reaction entropy:

$$\xi \, \Delta\hat{s}_r = \dot{\sigma}. \tag{b}$$

Using values from Example 5.7 and Example 4.19, we obtain the irreversibility for a time interval of eight hours as

$$I = N(CO_2) \, T \, \Delta\hat{s}_r = 0.6 \times 10^{-3} \times 298 \times 288.5 = 51.6\,\text{J}. \tag{c}$$

In comparison, the energy bound in this period is

$$\Delta U = \Delta t \, (\dot{Q}_s - \dot{Q}_1) = 8 \times 3600 \times 0.01 = 288\,\text{J}. \tag{d}$$

5.5   Electrochemical processes

Movement of charged matter from one electrical potential to another by a chemical pump is described in Section 4.6. Such a process can be illustrated by the transport of Na^+ ions from a lower potential in the intracellular space through a cell membrane to a higher potential in the extracellular space, as shown in Figure 4.15. Energy is supplied by coupling to hydrolysis of ATP according to reaction (4.74).

Use of the second law for this process requires knowledge of the entropy change for the transfer of ions from an electrolyte at one potential to another. This entropy change is solely associated with changes in the entropy of mixing that originates from differences in concentrations, and can be expressed, for ideal mixtures, by (3.101) or, more generally, as a result of activities from Table 3.3, as

$$\hat{s}_e - \hat{s}_i = \hat{R} \ln (c_e/c_i).$$

In this case, the quantitative description of the isothermal, stationary processes in the five control volumes of Figure 4.15 in the form of energy conservation, i.e. (4.76)–(4.80), can be supplemented by the following associated entropy balances:

$$\xi_1 \Delta\hat{s}_{r,G} = \dot{Q}_1/T + \dot{\sigma}_1 \text{ (CV)}_1, \tag{39}$$

$$-\xi_2 \Delta\hat{s}_{r,\text{ATP}} = \dot{Q}_2/T + \dot{\sigma}_2 \text{ (CV)}_2, \tag{40}$$

$$\xi_2' \Delta\hat{s}_{r,\text{ATP}} = \dot{Q}_2'/T + \dot{\sigma}_2' \text{ (CV)}_2', \tag{41}$$

$$-\xi_3 \hat{R} \ln (c_e/c_i) = \dot{Q}_3/T + \dot{\sigma}_3 \text{ (CV)}_3, \tag{42}$$

$$\xi_3' \hat{R} \ln (c_e/c_i) = \dot{Q}_3'/T + \dot{\sigma}_3' \text{ (CV)}_3'. \tag{43}$$

The sum of (39)–(43) gives, in analogy with (4.81), for stationarity,

$$\xi_1 \Delta\hat{s}_{r,G} = (\dot{Q}_1 + \dot{Q}_2 + \dot{Q}_2' + \dot{Q}_3 + \dot{Q}_3')/T + \dot{\sigma} \text{ (CV)}, \tag{44}$$

where $\dot{\sigma} = \dot{\sigma}_1 + \dot{\sigma}_2 + \dot{\sigma}_2' + \dot{\sigma}_3 + \dot{\sigma}_3'$, and where all $\dot{\sigma}_j > 0$. In this way, there are restrictions on each of the heat powers in (39)–(43).

Example 5.12

To justify assumptions (a) in Example 4.21 regarding the ratios of heat losses, we first show that the second law is fulfilled. This implies only that the processes are possible. Next, we derive general limits for these ratios.

Using the condition $\dot{\sigma}_j > 0$, numerical values for the reaction entropy of ATP-hydrolysis calculated in Example 5.10, and other values from Example 4.21, we derive conditions for $\dot{Q}_2$ and $\dot{Q}_3$ in (40) and (42):

$$-\dot{Q}_2 > \xi_2\, T\, \Delta \hat{s}_{r,\text{ATP}} = 13.7 \times 10^{-6} \times 310 \times 84 = 0.365\ \text{W}, \tag{a}$$

$$-\dot{Q}_3 > \xi_3\, \hat{R}T\, \ln(c_e/c_i) = (0.001/60) \times 8.314 \times 310 \times \ln(145/12)$$
$$= 0.107\ \text{W}. \tag{b}$$

According to (h) and (b) in Example 4.21:

$$-\dot{Q}_2 = 0.379\ \text{W}; \quad -\dot{Q}_3 = 0.145\ \text{W}, \tag{c}$$

so that (a) and (b) are fulfilled and the processes are possible. In addition to this result, it is possible in the present case, from the first and second law, and other information, to derive general expressions for the limitations of the distribution of the total heat losses in the individual processes in Figure 4.15. The sum of (4.76) and (4.77) can thus be rewritten as

$$\dot{Q}_2/\xi_2 = [(\xi_1/\xi_2)\, \Delta \hat{h}_{r,\text{G}} - \Delta \hat{h}_{r,\text{ATP}}]/(1 + \dot{Q}_1/\dot{Q}_2), \tag{d}$$

while (40), or (a), requires that

$$-\dot{Q}_2/\xi_2 \geqslant T\, \Delta \hat{s}_{r,\text{ATP}}. \tag{e}$$

The ratio of (d) and (e) gives the condition

$$\dot{Q}_1/\dot{Q}_2 \leqslant -[(\xi_1/\xi_2)\, \Delta \hat{h}_{r,\text{G}} - \Delta \hat{h}_{r,\text{ATP}}]/(T\, \Delta \hat{s}_{r,\text{ATP}}) - 1$$
$$\leqslant (2867/38 - 20)/26 - 1 \leqslant 1.13, \tag{f}$$

or $\dot{Q}_2/\dot{Q}_1 \geqslant 0.883$.

Correspondingly, the ratio of (4.80) and (43), or (b) rewritten as:

$$-\dot{Q}_3'/\xi_3' = z(\text{Na}^+)\mathscr{F}\,(\mathscr{E}_e - \mathscr{E}_i), \tag{g}$$

$$-\dot{Q}_3/\xi_3 > \hat{R}T\ln(c_e/c_i), \tag{h}$$

gives the condition

$$\dot{Q}_3/\dot{Q}_3' \geqslant \hat{R}T\ln(c_e/c_i)/[z(\text{Na}^+)\mathscr{F}\,(\mathscr{E}_e - \mathscr{E}_i)] = 0.74. \tag{i}$$

The difference between the assumptions in Example 4.21, $\dot{Q}_2/\dot{Q}_1 = 1$ and $\dot{Q}_3/\dot{Q}_3' = \text{4/5}$, and the above derived limit values of 0.883 and 0.74, respectively, expresses the degree of irreversibility of the processes, cp. Example 8.7.

Finally, note that, for the total process, (CV) in Figure 4.15, (44) gives

$$-\dot{Q} = \dot{E} = 1.032 \, \text{W} \geqslant -\xi_1 \, T \, \Delta \hat{s}_{r,G} = -0.360 \times 10^{-3} \times 63$$
$$= -0.023 \, \text{W (CV)}, \tag{j}$$

so that the second law is fulfilled.

5.6 Model of a biological system

A simplified model of a living system was described in Section 4.7 (Figure 4.16). It contains five representative basal processes (Table 4.2). They include, (1) oxidation of fuel to produce internal chemical energy supply (as ATP) for maintenance of other processes, (2) synthesis of AA from A, (3) synthesis of FF from F, (4) internal chemical pumping to 'lift' the non-structure L' to structure L'', and (5) generation of muscle tension T'' from T' to perform external work.

For the four last-mentioned processes, we can account for unsteadiness by the accumulation/deaccumulation in depots through the introduction of forward and backward reaction rates, ξ_i and ξ_i'. Whenever these rates are known, mole balances and energy conservation give a complete description of each process, including the energy expenditure of the total system and its parts, as derived in Section 4.7.

Corresponding to energy conservations (4.90)–(4.92), (4.96) and (4.98) for the five control volumes $(CV)_1$–$(CV)_5$ in Figure 5.4, (27) gives the following entropy balances:

$$\xi_1 \, \Delta \hat{s}_{r1} = \dot{Q}_1/T + \dot{\sigma}_1, \tag{45}$$

$$(\xi_2 - \xi_2') \, \Delta \hat{s}_{r2} = \dot{Q}_2/T + \dot{\sigma}_2, \tag{46}$$

$$(\xi_3 - \xi_3') \, \Delta \hat{s}_{r3} = \dot{Q}_3/T + \dot{\sigma}_3, \tag{47}$$

$$(\xi_4 - \xi_4') \, \Delta \hat{s}_{r4} = \dot{Q}_4/T + \dot{\sigma}_4, \tag{48}$$

$$(\xi_5 - \xi_5') \, \Delta \hat{s}_{r5} = \dot{Q}_5/T + \dot{\sigma}_5, \tag{49}$$

where, as noted earlier, the internal transfer of chemical energies $\dot{W}_{Ch2}$ to $\dot{W}_{Ch5}$ does not appear. The sum of (45)–(49) gives, for the whole system (CV),

$$\Sigma \, \xi_j \, \Delta \hat{s}_{rj} = \Sigma \, \dot{Q}_j/T + \dot{\sigma}, \tag{50}$$

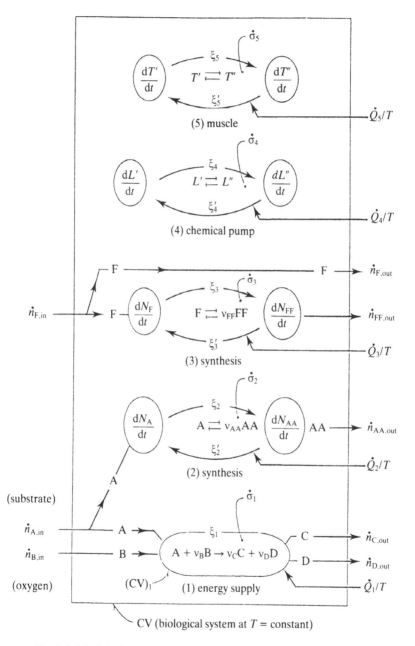

Fig 5.4 Model of a biological system at temperature T with (1) energy supply to operate (2) synthesis, (3) synthesis, (4) chemical pump and (5) muscle for internal and external mechanical work; d( )/dt symbolizes depot for accumulation/deaccumulation.

where ξ_j denotes the net reaction rate for the individual reaction. The second law requires that $\dot{\sigma}_j \geq 0$ and $\dot{\sigma} \geq 0$.

For special cases, we note stationarity for all processes, where $\xi_j = \xi_j'$, $j = 2,\ldots,5$ in (46)–(49), hence only the contribution for $j = 1$ will appear in the first summation of (50). This and other results, including expressions for the irreversibilities, will be developed later in relation to the concept of useful energy (Chapter 8).

6

Thermodynamic equilibrium

Although a certain departure from equilibrium is required for the occurrence of any process, the concept of thermodynamic equilibrium is of considerable importance. Thus, the concept is fundamental for the definition of state properties (Section 3.8) and for establishing unique relations between these, as shown in the equations of state expressed by observable quantities (Section 3.8.4) and in the Gibbs relation, the Euler form and the Gibbs–Duhem relation (Section 3.10.1). In addition, driving forces for processes, and thereby process rates, are often expressed in terms of deviations from equilibrium.

The fundamental definition of equilibrium, given in Section 3.8.1, refers to the isolated system, i.e. the closed system that does not exchange matter or energy with its surroundings. Assuming that all possible processes within such a system are allowed to proceed freely, the system is said to have reached an equilibrium state when these processes have ceased. Clearly, this definition can be applied to any mass in an open or closed system, since it is possible to imagine that the mass is suddenly isolated from its surroundings. If no processes then occur within the mass we conclude that its state was one of equilibrium.

Applying the first and the second laws to a simple, isolated system will show that the condition of thermodynamic equilibrium is equivalent to the system attaining a maximal value for its entropy. Equilibrium at conditions other than those of the isolated system leads to minima for the internal energy, the enthalpy, the Helmholtz function and the Gibbs function respectively.

Furthermore, the condition of entropy maximum for an isolated system implies that the equilibrium state has to be uniform with respect to the intensive properties temperature, pressure, and electrochemical potential for each component in the general case of a multicomponent compressible mixture. This is demonstrated by assuming the isolated system to consist of two subsystems without internal constraints for the mutual

163

exchange of energy and matter. Total equilibrium therefore consists of thermal, mechanical and electrochemical equilibrium.

6.1   Equilibrium in a simple system

We first consider a simple system which is assumed to be homogeneous. Non-homogeneous systems, e.g. consisting of several phases, are treated later as composite systems consisting of several subsystems.

The first law ((4.1)) and the second law ((5.1)) for a mixture of components that undergoes an arbitrary process in a control mass can be written as

$$d\bar{U}/dt = \dot{Q} + \dot{W}, \tag{1}$$

$$dS/dt \geq \dot{Q}/T, \tag{2}$$

where $\bar{U} = \Sigma\,(\hat{u}_j + \hat{\varphi}_j)N_j$ includes possible contributions from electrical potential energy, cp. (3.133), and where $S = \Sigma\,\hat{s}_jN_j$. Here, changes in kinetic and potential mechanical energy are neglected, cp. the note in the first part of Section 5.3. Elimination of $\dot{Q}$ between (1) and (2) gives

$$dS/dt - (1/T)\,d\bar{U}/dt + (1/T)\,\dot{W} \geq 0, \tag{3}$$

which, for a differential process, reduces to

$$dS - (1/T)\,d\bar{U} + (1/T)\delta W \geq 0, \tag{4}$$

a result that could also have been obtained by elimination of δQ among (2.5) and (2.6). Equation (3) or (4) is the starting point for the derivation of conditions for equilibrium.

6.1.1   Maximal entropy for an isolated system

For a control mass which is isolated from the surroundings, $\dot{Q} = \dot{W} = 0$, and $d\bar{U}/dt = 0$ according to (1). For a compressible mixture, for which $\delta W = -pd\mathcal{V}$, the condition $\delta W = 0$ implies that $\mathcal{V} = $ constant. In this case, (3) reduces to

$$(dS/dt)_{\bar{U},\mathcal{V}} \geq 0 \text{ (towards equilibrium)}, \tag{5}$$

or, for a differential process,

$$dS_{\bar{U},\mathcal{V}} \geq 0 \text{ (towards equilibrium)}. \tag{6}$$

This implies that the entropy of the isolated system increases as long as there are spontaneous processes within the control mass, and that the entropy for fixed $\bar{U}, \mathcal{V}$ attains a maximal value at equilibrium:

$$S_{\bar{U}, \mathcal{V}} = S_{max} \text{ (equilibrium).} \tag{7}$$

These considerations formalize the concept of equilibrium.

For more general media than the simple compressible medium, the internal energy and entropy depend not only on the volume but also on a number of other extensive state properties X_j, in association with which energy can be supplied as work (cp. Table 3.2 in Section 3.5). The condition of isolation then implies fixed U, X_j. At present, we limit the treatment to simple mixtures. When these are incompressible, the volume is no longer a state property, and only U is fixed in (7). Since any state property is a unique function of other independent state properties at equilibrium, in accordance with the general relation (2.3), the statement of entropy maximum at equilibrium can also be expressed as follows. The equilibrium value of any state property which is not restricted by internal constraints (see later) is precisely the value which maximizes the entropy of the isolated system. This statement can be understood as a fundamental axiom and it contains the second law of thermodynamics (Callen, 1960).

6.1.2 Alternative conditions of equilibrium

Since (4) is general we may derive other conditions for equilibrium by requiring properties other than $U, \mathcal{V}$ to be kept constant while spontaneous processes are allowed to proceed to termination. Thus assuming the entropy and the volume to be constant in (4) yields

$$d\bar{U}_{S, \mathcal{V}} \leq 0 \text{ (towards equilibrium).} \tag{8}$$

This means that the electrochemical energy $\bar{U}$ of the system will decrease as long as spontaneous processes take place, and that $\bar{U}$ for fixed $S, \mathcal{V}$ attains a minimal value at equilibrium:

$$\bar{U}_{S, \mathcal{V}} = \bar{U}_{min} \text{ (equilibrium).} \tag{9}$$

The statement of the minimum of internal energy at equilibrium can be expressed in the following way. The equilibrium value of any state property, e.g. X_j, which is not restricted by internal constraints, is precisely the value that minimizes the internal energy of the system at fixed values of entropy and volume. This statement is well known for adiabatic, non-dissipative systems, e.g. in mechanics. Here, equilibrium states of deformation are determined as the stationary states that minimize the energy of deformation.

Other conditions for equilibrium can be derived from (4) by introducing the useful, derived state properties enthalpy, the Helmholtz function and the Gibbs function, and their differentials:

$$H \equiv U + p\,\mathcal{V}; \; \mathrm{d}H = \mathrm{d}U + p\,\mathrm{d}\mathcal{V} + \mathcal{V}\,\mathrm{d}p, \tag{10}$$

$$A \equiv U - TS; \; \mathrm{d}A = \mathrm{d}U - T\,\mathrm{d}S - S\,\mathrm{d}T, \tag{11}$$

$$G \equiv H - TS; \; \mathrm{d}G = \mathrm{d}U + p\,\mathrm{d}\mathcal{V} + \mathcal{V}\,\mathrm{d}p - T\,\mathrm{d}S - S\,\mathrm{d}T. \tag{12}$$

Thus, (12) substituted into (4), with $\delta W = -p\,\mathrm{d}\mathcal{V}$ for a compressible substance, gives

$$-S\,\mathrm{d}T - \mathrm{d}G + \mathcal{V}\,\mathrm{d}p \geqslant 0 \text{ (towards equilibrium)}, \tag{13}$$

from which, at fixed temperature and pressure,

$$\mathrm{d}G_{T,p} \leqslant 0 \text{ (towards equilibrium)}. \tag{14}$$

This implies that the Gibbs function of the system will decrease as long as spontaneous processes take place in the system, and that the Gibbs function for fixed T,p will attain a minimal value at equilibrium,

$$G_{T,p} = G_{\min} \text{ (equilibrium)}. \tag{15}$$

This, and other conditions for equilibrium are summarized in Table 6.1. The table shows a systematic relation between dependent variables and associated 'natural' independent variables: S and $\mathcal{V}$ are thus the natural variables for U. The derived functions A and G are particularly useful in that they provide expressions for equilibrium conditions at fixed observable quantities, $T,\mathcal{V}$ and T,p respectively.

Table 6.1. *Alternative conditions for equilibrium for a compressible medium*

Equilibrium criterion	Properties fixed
$S = S_{\max}$	$U, \mathcal{V}$
$U = U_{\min}$	$S, \mathcal{V}$
$H = H_{\min}$	S, p
$A = A_{\min}$	$T, \mathcal{V}$
$G = G_{\min}$	T, p

6.2   Equilibrium between subsystems

We now consider an isolated composite system C, consisting of two subsystems A and B, each of which is assumed be in equilibrium (Figure 6.1). Since the extensive state properties entropy, electrochemical energy, volume and mole number for each component are additive, we have

$$S_C = S_A + S_B, \tag{16}$$

$$\tilde{U}_C = \tilde{U}_A + \tilde{U}_B \; ; \; \mathscr{V}_C = \mathscr{V}_A + \mathscr{V}_B; \quad N_{jC} = N_{jA} + N_{jB}. \tag{17}$$

Condition (6) for equilibrium in the composite system C is expressed by using (16) as

$$dS_C = dS_A + dS_B \geq 0 \text{ (towards equilibrium)}, \tag{18}$$

subject to fixed $\tilde{U}_C$, $\mathscr{V}_C$ and N_C which, in accordance with (17), can be written as

$$d\tilde{U}_A + d\tilde{U}_B = 0; \; d\mathscr{V}_A + d\mathscr{V}_B = 0; \; dN_{jA} + dN_{jB} = 0. \tag{19}$$

We now assume that A and B are separately at equilibrium and at uniform state. Then, for each subsystem, the Gibbs relation (3.38) is valid and can be rewritten as

$$dS_A = (1/T_A) \, d\tilde{U}_A + (p_A/T_A) \, d\mathscr{V}_A - \Sigma \, (\bar{\mu}_{jA}/T_A) \, dN_{jA}, \tag{20}$$

$$dS_B = (1/T_B) \, d\tilde{U}_B + (p_B/T_B) \, d\mathscr{V}_B - \Sigma \, (\bar{\mu}_{jB}/T_B) \, dN_{jB}. \tag{21}$$

When these expressions, and (19), are inserted into (18), we obtain

$$\begin{aligned} dS_C = (1/T_A - 1/T_B) \, d\tilde{U}_A + (p_A/T_A - p_B/T_B) \, d\mathscr{V}_A \\ - \Sigma \, (\bar{\mu}_{jA}/T_A - \bar{\mu}_{jB}/T_B) \, dN_{jA} \geq 0, \end{aligned} \tag{22}$$

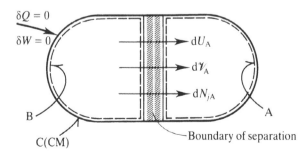

Fig 6.1 Isolated composite system C, consisting of subsystems A and B.

where the summation in the last term includes the k components $j = 1,2,\ldots,k$ of the mixture.

In order to understand the meaning of this important result, we need to clarify the concept of an *internal constraint*. We say that there are no internal constraints in system C when the imaginary or real boundary between subsystems A and B allows free flow of energy (e.g. by way of heat conduction through the boundary), free flow of volume (e.g. by the displacement of the boundary as a piston with the transfer of energy as work), and free flow of matter of component j (e.g. by the boundary being permeable with respect to component j). An internal constraint on one of the aforementioned quantities means that this quantity cannot be exchanged, and that equilibrium with respect to this quantity cannot be established.

When there are no internal constraints in C, S_C will increase to a maximal value $S_{C,max}$, cp. (7). This value is obtained when all processes have ceased so that $dS_C = 0$ in (22), and the system is at equilibrium. For small arbitrary displacements, e.g. $dU_A \neq 0$, etc., close to this state, the condition $dS_C = 0$ in (22) requires that

$$T_A = T_B \text{ (thermal equilibrium)}, \tag{23}$$

$$p_A = p_B \text{ (mechanical equilibrium)}, \tag{24}$$

$$\bar{\mu}_{jA} = \bar{\mu}_{jB} \text{ (electrochemical equilibrium)}. \tag{25}$$

While (23) and (24) are readily understood, (25) is not part of our direct experience.

> In the case where the boundary between A and B is heat conducting, but immovable and impermeable for all components j, i.e. C has internal constraints $d\mathcal{V}_A = 0$ and $dN_{jA} = 0$, (22) shows that thermal equilibrium but not mechanical or electrochemical equilibrium, can be established between the subsystems. The situation is different in the case of an adiabatic and impermeable, but frictionless, movable piston as boundary. Using internal constraints $dN_{jA} = 0$ in (22), no answer is obtained. The expected solution $p_A = p_B$ is obtained only by use of the first law and conditions (19).

In addition to conditions for equilibrium, (22) contains the second law and can therefore also be used to show the direction of irreversible processes. For example, if $T_A \neq T_B$ initially, we have, for $dU_A \neq 0$, that

$$dS_c = (1/T_A - 1/T_B)\, dU_A \geq 0. \tag{26}$$

Hence, $T_B > T_A$ would imply $dU_A > 0$. Now, the first law for system A gives $dU_A = \delta Q_{B \to A}$, so the result is in agreement with our experience.

Energy as heat is transferred from the system with a higher temperature to the system with a lower temperature. In the same way, an inequality in pressure will displace a movable piston in such a direction that the pressure difference disappears. These two processes are examples of the *Le Chatelier principle*, which is discussed for chemical reactions in Section 6.5.

6.3 Thermomechanical equilibrium

For simple compressible systems, involving no chemical reactions, (22) requires that temperature and pressure be uniform at thermodynamic equilibrium. This is shown in Example 6.1 below, even though such equilibria are of limited importance for biological systems.

Note that the condition for mechanical equilibrium can be obtained from Newton's second law ((2.9)), which shows that when the acceleration is zero, the sum of forces becomes zero for the system considered. This fact was used in the discussion of the more complicated problem of lung alveoli in Section 3.5.2 in order to show that the excess pressure Δp in a spherical gas bubble of radius R is

$$\Delta p = 2\,\sigma_s/R \tag{27}$$

where σ_s is the surface tension associated with the gas–liquid interface. However, result (27) can also be derived from (4). For a gas bubble, including its gas–liquid interface, the reversible work associated with a differential change in its volume, and hence its surface area (cp. Table 3.2 in Section 3.5.1), is

$$\delta W = \sigma_s\,dA - \Delta p\,d\mathcal{V} = (2\sigma_s/R - \Delta p)\,d\mathcal{V}, \tag{28}$$

where the relations $A = 4\pi R^2$ and $\mathcal{V} = \tfrac{4}{3}\pi R^3$ for a sphere have been used. If the two first terms in (4) are zero (which is true for an isolated system) it follows that $\delta W = 0$ at equilibrium. This result, and an arbitrary displacement $d\mathcal{V} \neq 0$, shows that (28) reduces to (27).

Example 6.1

In a closed cylinder isolated from the surroundings, a fixed piston represents the boundary between subsystems A and B (cp. the schematic Figure 6.1). Initially, the subsystems have the same volume $\mathcal{V}_A = \mathcal{V}_B = 1.0\,l$ and contain atmospheric air at the same temperature $T_A = T_B = 300$ K, but at different pressures, $p_A = 2$ bar and $p_B = 1$ bar.

From this initial state the piston is released. It is assumed to be heat conducting, freely movable and impermeable to gases. We seek the volume, pressure and temperature for the air in A and B when equilibrium has been attained and the piston no longer moves.

The air is treated as an ideal gas conforming to the equation of state (3.50), so that the initial state is given by

$$p_A \mathcal{V}_A = N_A \hat{R} T_A; \quad p_B \mathcal{V}_A = N_B \hat{R} T_B. \tag{a}$$

In the final state, denoted by ( )', thermal and mechanical equilibrium ((23) and (24)) prescribes

$$p'_A \mathcal{V}'_A = N_A \hat{R} T'_A; \quad p'_B \mathcal{V}'_B = N_B \hat{R} T'_A, \tag{b}$$

and the total volume is constant:

$$\mathcal{V}'_A + \mathcal{V}'_B = 2 \mathcal{V}_A. \tag{c}$$

We complete the formulation of the problem by writing the first law ((4.2)), the right-hand side of which is zero for system C:

$$(U'_A + U'_B) - (U_A + U_B) = 0, \tag{d}$$

or using (3.56) for an ideal gas and $\hat{c}_v =$ constant:

$$N_A \, \hat{c}_v(T'_A - T_A) + N_B \, \hat{c}_v(T'_B - T_B) = 0. \tag{e}$$

Since $T'_A = T'_B$, $T_A = T_B$ and $N_A \neq N_B$, we conclude that $T'_A = T_A$. This is not surprising since the internal energy of an ideal gas depends on the absolute temperature only and since the initial and the final states are isothermal. Now, (a) and (b) give

$$p_A/p_B = N_A/N_B = \mathcal{V}'_A/\mathcal{V}'_B, \tag{f}$$

which with the information stated and (c) gives

$$\mathcal{V}'_A = \tfrac{4}{3} \text{ litre}; \quad \mathcal{V}'_B = \tfrac{2}{3} \text{ litre}; \quad p'_A = p'_B = \tfrac{3}{2} \text{ bar}, \tag{g}$$

and $T'_A = T'_B = 300$ K.

Applying the first law to each of the subsystems shows that energy is exchanged as both heat and work so that the internal energy remains constant. Use of the second law for the total system shows, as expected, that the process is irreversible.

6.4   Non-reactive electrochemical equilibrium

In this section we consider equilibrium states for compressible media with several phases that are in contact along plane boundaries (so that surface tensions can be neglected). A boundary can be a gas–liquid interface, which normally has no constraints, or a membrane, which normally has constraints with respect to the transfer of some of the components in a mixture.

6.4.1   Phase equilibria. The Gibbs phase rule

Consider an isolated multicomponent system consisting of several phases, A, B, C,. . . in mutual equilibrium and without internal constraints. The assumptions in Section 6.2 will then lead to a state of uniform temperature (thermal equilibrium):

$$T_A = T_B = T_C = \ldots = T, \tag{29}$$

or uniform pressure (mechanical equilibrium),

$$p_A = p_B = p_C = \ldots = p, \tag{30}$$

and having the same value for the chemical potential of any given component in all phases (chemical equilibrium with no differences in electrical potential):

$$\left.\begin{array}{l} \hat{\mu}_{jA}(T,p,\chi_{iA}) = \hat{\mu}_{jB}(T,p,\chi_{iB}) = \hat{\mu}_{jC}(T,p,\chi_{iC}) = \ldots, \\ j = 1,2, \ldots, k; i = 1,2 \ldots, k-1. \end{array}\right\} \tag{31}$$

Expression (31) implies that the chemical potential of each of the k components in each phase depends on the temperature, pressure and composition of that phase. Note that the specific composition is given by $k-1$ mole fractions χ_i, cp. (3.79) and (3.63).

If p denotes the number of phases and k the number of components, the thermodynamic state of the composite system is uniquely determined by the temperature T, pressure p, and $k-1$ mole fractions in each of the p phases. There is thus a total of $2 + p(k-1)$ variables. These variables are constrained by p-1 equilibrium conditions (31) for each of the k components, i.e. a total of $(p-1)k$ equilibrium relations. The difference between the two expressions gives the number of independent variables f that can be specified freely to determine a unique equilibrium state:

$$f = k - p + 2, \tag{32}$$

which is the *Gibbs phase rule*.

Example 6.2

For a gas, e.g. water vapour, (32) gives $f = 1 - 1 + 2 = 2$, so that two variables, e.g. T and p, determine the state.

For boiling water in equilibrium with water vapour, only the temperature or the pressure can be specified, since (32) gives $f = 1 - 2 + 2 = 1$.

At the triple point, where solid, liquid and gaseous phases are present, (32) gives $f = 1 - 3 + 2 = 0$ for one component.

The equilibrium state for a two component mixture present in two phases is determined by $f = 2 - 2 + 2 = 2$ independent parameters, e.g. the pressure of the mixture and the specific composition of one of the phases.

6.4.2 One-component phase equilibrium. The Clapeyron equation

As an example of the use of the condition for phase equilibrium, we derive an expression for the pressure dependence of the boiling point of a liquid. At equilibrium between the two phases A and B, consisting of one and the same component, (31) gives at T,p

$$\hat{\mu}_A(T,p) = \hat{\mu}_B(T,p) \quad \text{or} \quad \hat{g}_A = \hat{g}_B, \tag{33}$$

since the chemical potential of a pure substance is identical to the Gibbs function. Hence (33) can be written as

$$\hat{h}_B - \hat{h}_A = T(\hat{s}_B - \hat{s}_A) \tag{34}$$

so that changes in enthalpy and entropy at a change of phase are simply related.

Introducing the Gibbs relation as expressed by $\hat{g}$,

$$\mathrm{d}\hat{g} = -\hat{s}\,\mathrm{d}T + \hat{v}\,\mathrm{d}p, \tag{35}$$

into the differential of (33), $d\hat{g}_A = d\hat{g}_B$, and using (34) we obtain the *Clapeyron equation*

$$(dp/dT)_{AB} = (\hat{h}_B - \hat{h}_A)/[T(\hat{v}_B - \hat{v}_A)]. \tag{36}$$

This gives the relation between changes in pressure and temperature at phase equilibrium, i.e. the slope in a p–T diagram of curves of melting, sublimation or evaporation. For the two last-mentioned cases, (36) can be simplified by assuming the vapour phase (e.g. B) to be an ideal gas, $p\hat{v}_B = \hat{R}T$, and we assume that $\hat{v}_B \gg \hat{v}_A$. This gives

$$(dp/dT)_{AB} \approx p(\hat{h}_B - \hat{h}_A)/(\hat{R}T^2). \tag{37}$$

6.4.3 Two-component phase equilibrium

The description of phase equilibrium in multicomponent mixtures is both theoretically and experimentally difficult (Rock and Gerholt, 1974; Prigogine and Defay, 1954; Prausnitz, 1969). The number of variable thermodynamic state properties is often large, and most mixtures are non-ideal. We limit treatment here to two simple cases of binary systems, ideal mixtures and dilute solutions.

We consider two homogeneous phases (Figure 6.2(a)) where the composition of the liquid phase (A) is given by the mole fractions χ_1 and $\chi_2 = 1 - \chi_1$ and where the composition of the vapour phase (B) is given by the partial pressures p_1 and $p_2 = p - p_1$. At equilibrium, (32) shows, using $k = 2$ and $p = 2$, that $f = 2$. Therefore, two independent thermodynamic state properties fix the specific state in this case, for example T and χ_2 (Figure 6.2(b)), or T and p. At thermal and mechanical equilibrium, fulfilled by uniform temperature T and pressure p, the two unknowns are determined by condition (25) for chemical equilibrium:

$$\hat{\mu}_{1A} = \hat{\mu}_{1B} \; ; \; \hat{\mu}_{2A} = \hat{\mu}_{2B}. \tag{38}$$

First, consider the case of ideal mixtures. From (3.122) and (3.124) using $f_j = 1$ for the liquid phase and using partial pressures for the gas phase, cp. (3.110), (38) gives *Raoult's law*:

$$p_1 = \chi_1 \, p_1^*(T); \quad p_2 = (1 - \chi_1) \, p_2^* \, (T), \tag{39}$$

where $p_j^*(T)$ denotes the vapour pressure of the pure component j at temperature T. Result (39), shown graphically in Figure 6.2(b), expresses that the partial pressure of a component in the vapour phase is the fraction of the vapour pressure of the pure component that is given by the

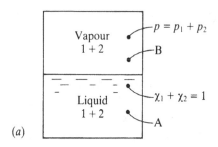

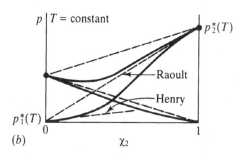

Fig 6.2 Binary mixture in liquid phase (A) and vapour phase (B) in phase equilibrium at p,T. (a) Notation. (b) Partial pressures of components and total pressure ($p = p_1 + p_2$) as a function of composition of the liquid phase at constant T: – – – ideal mixtures (Raoult's law); ———— non-ideal mixtures (with Henry's law shown for dilute solutions).

mole fraction in the liquid phase. The composition of the vapour phase expressed in terms of T,p is obtained from (39) by elimination of χ_1 and use of $p = p_1 + p_2$:

$$p_1 = [p_2^*(T) - p]/[p_2^*(T)/p_1^*(T) - 1]. \tag{40}$$

The composition of the liquid phase is then determined by (39).

Next, consider the case of a dilute, non-ideal solution of a solute, e.g. $\chi_2 \ll 1$, in a solvent $j = 1$. Here, Raoult's law (39) can be used for the solvent:

$$p_1 \approx \chi_1 p_1^*(T) = (1 - \chi_2) p_1^*(T), \tag{41}$$

while *Henry's law* may be used for the solute:

$$p_2 \approx \chi_2 K_2, \tag{42}$$

where Henry's constant K_2 is determined by experiment. Result (42) describes the *solubility* of a gas component in terms of its mole fraction in a liquid when the partial pressure p_2 in the gas phase is given. Other forms of (42) involve concentration or volume fraction, and the proportionality constant is denoted the *solubility coefficient*, the unit of which specifies the quantities involved.

Finally, it should be noted that when the difference is large between the saturation pressures of the two components, e.g. $p_2^* \gg p_1^*$, (40) and (39) show, for an ideal gas phase, that the liquid phase is a dilute solution of the more volatile component 2, i.e. $\chi_1 \approx 1$. So, neglecting the solubility of one of the components, (41) shows that its composition at equilibrium is determined by the vapour pressure of the liquid at the temperature in question.

Example 6.3

The partial pressure of oxygen in dry atmospheric air at $p = 1$ atm is $0.21 \times 760 = 160$ mm Hg. In air saturated with water, the partial pressure of water vapour approximately equals the water vapour pressure at $37\,°C$, $p^* = 47$ mm Hg, see discussion following (42). The partial pressure of oxygen here is $0.21 \times (760 - 47) = 150$ mm Hg. In the lung alveoli, the partial pressure of oxygen is lower, $p_2 \approx 100$ mm Hg, because oxygen is continuously taken up by the blood to be consumed in the body. The velocity of the blood flow in the walls of the alveoli is small under normal resting conditions so there is practically equilibrium between the alveolar air and the blood leaving the alveoli. Here, any membrane effect on the equilibrium is ignored.

Calculate the content of physically dissolved oxygen in the blood leaving the alveoli when the solubility coefficient is $\lambda = 22$ ml (STPD) per litre of blood and atm. at $38\,°C$.

Using the equivalent of (42) and $\mathcal{V}_{STP} = 22.4$ l for one mole dry air from (a) in Example 3.16, we obtain

$$\mathcal{V}_2/\mathcal{V} = \lambda p_2 = 22.0 \times (100/760) = 2.90 \text{ mol/l} = 2.90/22.4$$
$$= 0.13 \text{ mmol/l}. \tag{a}$$

In reality, the dissolved oxygen is in equilibrium with the haemoglobin molecules in the blood, and the oxygen content is much larger, about 10 mmol/l.

6.4.4   Membrane equilibrium

An important class of biological processes involves the transport of matter across membranes. Due to their physical and biochemical structure, membranes restrict these processes by having different permeabilities for different components in the mixtures that are kept apart by the membrane. A closer understanding of these phenomena is gained by first studying the equilibrium situations across and within a membrane and determining the state of the mixtures.

For biological systems, it is sufficient to begin the analysis with a composite isothermal system at temperature T consisting of two subsystems A and B separated by a non-movable membrane, cp. Figure 6.1 for $d\mathcal{V}_A = 0$. At equilibrium, we have thermal equilibrium ((23)), which is generally fulfilled for biological systems, and electrochemical equilibrium ((25)) for some but not all components in the mixtures. In addition, the principle of electroneutrality is satisfied. The missing mechanical equilibrium ((24)) implies a pressure difference (*osmotic pressure*) between subsystems A and B. The existence of both non-permeating and permeating ions implies an electrical potential difference (the *membrane potential*) across the membrane.

Introducing activities in accordance with (3.136) and considering an isothermal system with $T_A = T_B = T$, condition (25) for electrochemical equilibrium of a permeating component j can be written generally as

$$[\hat{\mu}_j^0 + \hat{R}T \ln (a_j/a_j^0) + z_j \mathscr{FE}]_A = [\hat{\mu}_j^0 + \hat{R}T \, n \, (a_j/a_j^0) + z_j \mathscr{FE}]_B. \quad (43)$$

Here, the chemical potential at the reference state depends, in general, on the state of the mixture, but for ideal mixtures it can be expressed by the Gibbs function of the component $\hat{\mu}_{jA}^0 = \hat{g}_{jA}(T,p_A)$, cp. Section 3.10.5. The activity units and standard activities in (43) are normally chosen according to Table 3.3.

Many transport processes and chemical reactions in biological systems take place in phases which are dilute aqueous solutions. For a large number of components, including water, it is therefore reasonable to assume that the standard activity for each component and the chemical standard potential for each component assume the same value in such phases at the same temperature and pressure, i.e.

$$a_{jA}^0 = a_{jB}^0 \quad \text{at} \quad (T,p)_A = (T,p)_B.$$

It should be noted that this approximation has been questioned after realizing that many important chemical reactions in the cells take place in so-called multienzyme systems that are associated with intracellular structures that are more or less solid (see, e.g., Welch, 1985).

If the pressures in A and B are different, (35) is integrated for permeating component j between p_A and p_B. Since the temperature is constant, and the molar volume of the component as pure substance is assumed to be constant, cp. (3.131) and (3.132), we obtain

$$\hat{\mu}_{jB}^0 - \hat{\mu}_{jA}^0 = \hat{g}_{jB}^0 - \hat{g}_{jA}^0 = \hat{v}_j(p_B - p_A). \tag{44}$$

Inserting (44) into (43) and using $a_{jA}^0 = a_{jB}^0$ gives the general expression for *membrane equilibrium*:

$$\hat{v}_j(p_B - p_A) + \hat{R}T \ln (a_{jB}/a_{jA}) + z_j \mathscr{F} (\mathscr{E}_B - \mathscr{E}_A) = 0. \tag{45}$$

Substituting activity coefficients according to (3.124), we obtain

$$\hat{v}_j(p_B - p_A) + \hat{R}T \ln [\chi_j/f_j)_B/(\chi_j/f_j)_A] + z_j \mathscr{F} (\mathscr{E}_B - \mathscr{E}_A) = 0, \tag{46}$$

or by introducing concentrations according to (3.125):

$$\hat{v}_j(p_B - p_A) + \hat{R}T \ln [(c_j/y_j)_B/(c_j/y_j)_A] + z_j \mathscr{F} (\mathscr{E}_B - \mathscr{E}_A) = 0, \tag{47}$$

where $f_j = 1$ and $y_j = 1$ under ideal conditions, or, less restrictively, $f_{jA}/f_{jB} = 1$ and $y_{jA}/y_{jB} = 1$. If, furthermore, $p_B = p_A$, (47) reduces to the *Nernst equation*,

$$\Delta \mathscr{E} = \mathscr{E}_B - \mathscr{E}_A = - (\hat{R}T/z_j \mathscr{F}) \ln [(c_j/y_j)_B/(c_j/y_j)_A].$$

When the equilibrium condition (45), (46) or (47) is used, together with the condition of electroneutrality, it describes the so-called *Donnan* (Gibbs–Donnan) *equilibrium* across a membrane that separates solutions containing non-permeating ions (see Example 6.6). In this case, differences will appear in both pressure and electrical potential. However, if the non-permeating components are electrically neutral, only the pressure difference appears. The practical use of (45)–(47) for determining osmotic pressure difference and membrane potential will be illustrated in the examples that follow. First, however, we describe an approximation that is valid for dilute solutions, and identify the number of variables that determine the equilibrium state.

In biological systems, solutions are normally diluted and certain simplifications can be made. The equilibrium condition (45) is valid for each permeating component, but if this is an ion, the first term in (45) is negligible in comparison with the other two terms and the activity difference determines the membrane potential. For water, the last term in (45) disappears since $z_j = 0$, and the activity difference determines the osmotic pressure difference. When calculating this pressure difference, we can often introduce the following approximations. First, for the ideal case, $f_j = 1$ in (46), we obtain

$$\pi \equiv p_B - p_A = - (\hat{R}T/\hat{v}_w) \ln (\chi_{wB}/\chi_{wA}). \tag{48}$$

Next, using the definitions in Section 3.9.1,

$$\chi_{wA} = 1 - \sum_2^k \chi_{iA}; \quad \hat{v}_w = \mathcal{V}/N_w; \quad c_{iA} = N_{iA}/\mathcal{V},$$

and the approximation $\ln (1 - x) \approx -x$ for $x \ll 1$, we obtain

$$(1/\hat{v}_w) \ln \chi_{wA} \approx -(N_w/\mathcal{V}) \sum_2^k \chi_{iA}$$

$$= -(N_w/\mathcal{V}) \sum_2^k N_{iA}/N \approx - \sum_2^k c_{iA}, \tag{49}$$

since $N_w \approx N$ for a dilute solution. Substituting (49), and a corresponding expression for $\ln \chi_{wB}$, into (48), we obtain the approximation, denoted the *van't Hoff equation*:

$$\pi = p_B - p_A \approx \hat{R}T \left[\sum_2^k c_{iB} - \sum_2^k c_{iA} \right]. \tag{50}$$

In a system with one single non-permeating component dissolved in water, $k = 2$ and (50) reduces to $\pi = \hat{R}T(c_{2B} - c_{2A})$. If component 2 is present in phase B only, we obtain $\pi = \hat{R}Tc_{2B}$.

If $p = 2$ denotes the number of phases and m the number of permeating components with known ionic valencies z_j, the thermodynamic state of the composite system is determined uniquely by the quantities temperature T, pressures p_A and p_B, mole fractions χ_j in each of the two phases, and the electrical potential difference $\mathcal{E}_B - \mathcal{E}_A$ across the membrane, i.e. a total of $1 + 2 + 2m + 1 = 4 + 2m$ variables. These variables are restricted by m equilibrium relations ((43)). The difference between the two expressions determines the number of independent variables, f, that can be freely specified for a unique state:

$$f = 4 + m, \tag{51}$$

which is a special form of the Gibbs phase rule for electrochemical or chemical membrane equilibrium.

Example 6.4

For the definition of osmotic pressure in Section 3.10.5, illustrated by Figure 3.13, $m = 1$ for a single permeating component, e.g. water, and $f = 5$ according to (51).

With reference to Example 3.22, the state is uniquely defined by specifying the salt concentration in phase A (0%) and in phase B (2.5%), the temperature (300 K), the pressure in one phase (e.g. p_A) and the electrical potential difference across the membrane (in this case $= 0$), a total of five quantities in accordance with (51). The equilibrium condition ((43)), which reduces to (3.130) or (46), determines the pressure difference $p_B - p_A$, which is the osmotic pressure of phase B relative to phase A.

Analogous reasoning applies to an ideal gas mixture separated by a semi-permeable membrane, see Example 3.23, where it is shown that the partial pressure of the permeating gas component must be the same on both sides of the membrane at equilibrium.

Example 6.5

A solution of $c_{2B} = 0.1$ mole per litre of sucrose in water (phase B) is in equilibrium with pure water (phase A) across a membrane which is impermeable to sucrose. We wish to calculate the hydrostatic pressure difference (the osmotic pressure) that exists at equilibrium in the case where the solution is ideal and the temperature is 37 °C in both phases.

Since the electrical potential difference is zero, $f_j = 1$ for the ideal case, and component $j = 1$ refers to water as the only permeating compound and $j = 2$ to sucrose, (46) gives (48), or, as an approximation, (50), which with values in MKS units gives

$$\pi = \hat{R}Tc_B = 8.314 \times 310 \times 0.1 \times 10 = 257.7 \,\text{kPa}. \tag{a}$$

It is often convenient to use the alternative value from (3.52), $R = 0.082\,07$ atm-l/mol-K:

$$\pi = RTc_{2B} = 0.082\,07 \times 310 \times 0.1 = 2.54 \,\text{atm}. \tag{b}$$

Example 6.6

An aqueous solution (phase A) of 100 mmol/l of NaCl is in equilibrium across a protein-tight membrane with an aqueous solution (phase B) of NaCl and protein. The protein concentration is 5 mmol/l with a negative ionic valency of 10.

We seek the differences in electrical potential and hydrostatic pressure across the membrane when both solutions are assumed to be ideal and the temperature is 25 °C (Figure 6.3).

There is a total of four components, water (w), sodium ions (Na), chloride ions (Cl) and protein (P), of which the first three are permeating. The equilibrium condition ((47)) with $y_{jA}/y_{jB} = 1$ for sodium and chloride ions, and after neglecting the first term, (47) gives

$$\hat{R}T \ln \left[c(\text{Na})_B / c(\text{Na})_A \right] + z(\text{Na}) \mathscr{F} \left(\mathscr{E}_B - \mathscr{E}_A \right) = 0, \tag{a}$$

$$\hat{R}T \ln \left[c(\text{Cl})_B / c(\text{Cl})_A \right] + z(\text{Cl}) \mathscr{F} \left(\mathscr{E}_B - \mathscr{E}_A \right) = 0, \tag{b}$$

while ((46)) for water, where $z = 0$, gives

$$\hat{v}_w(p_B - p_A) + \hat{R}T \ln \left[\chi_{wB} / \chi_{wA} \right] = 0. \tag{c}$$

Finally, the condition of electroneutrality in phase B gives

$$\left[(zc)_{\text{Na}} + (zc)_{\text{Cl}} + (zc)_P \right]_B = 0, \tag{d}$$

or, with $z(\text{Na}) = +1$, $z(\text{Cl}) = -1$ and $z(\text{P}) = -10$,

$$\left[c(\text{Na}) - c(\text{Cl}) - 10c(\text{P}) \right]_B = 0, \tag{e}$$

so that the concentration of sodium and chloride ions cannot be the same in phase B. The condition of electroneutrality in phase A is assumed to be fulfilled, i.e. $c(\text{Na}) = c(\text{Cl}) = 100$ mmol/l, which is used in (f) below.

Relations (a), (b), (c) and (e) determine, with the given data, differences in electrical potential and pressure across the membrane, and the concentration of sodium and chloride ions in phase B. The sum of (a) and (b) gives, with the stated concentrations in phase A,

$$\left[c(\text{Na}) \, c(\text{Cl}) \right]_B = \left[c(\text{Na}) \, c(\text{Cl}) \right]_A = 100 \times 100. \tag{f}$$

When $c(\text{Na})_B$ is eliminated between (e) and (f), we obtain a second-order equation with one positive root for $c(\text{Cl})_B$, so that $c(\text{Na})_B$ can also be determined:

$$c(\text{Cl})_B = 78 \text{ mmol/l}; \quad c(\text{Na})_B = 128 \text{ mmol/l}. \tag{g}$$

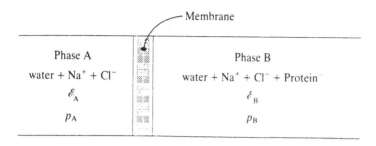

Fig 6.3 Membrane equilibrium, Example 6.6.

The electrical potential difference is now obtained from (a) (or b):

$$\mathcal{E}_B - \mathcal{E}_A = -[8.314 \times 298/(1 \times 96\,500)] \ln (128/100) = -6.4\,\text{mV}, \tag{h}$$

which shows the general result that the algebraic sign of the potential difference is determined by the algebraic sign of the charge of the non-permeating ion.

The pressure difference is calculated from (c), which is approximated by (50):

$$\pi = p_B - p_A \approx 8.314 \times 298 \times [(128 + 78 + 5) - (100 + 100)]$$
$$= 27.3\,\text{kPa} = 0.27\,\text{atm}, \tag{i}$$

which shows the general result that the pressure is largest on the side of the membrane having the largest concentration of non-permeating components. In the absence of NaCl, the potential difference disappears, and the pressure difference decreases, in this case to about one half of the value found above.

While the relatively modest pressure difference of 0.27 atm can be sustained by a membrane in a laboratory experiment, this is impossible for a cell membrane in a living animal. Here, a very slight pressure difference will activate transport processes in the membrane which will effectively eliminate the pressure difference. Such transport often takes place through the operation of chemical pumps which move sodium ions from the protein phase to the aqueous phase. We can illustrate the relatively small changes in the concentration that are necessary to eliminate the osmotic pressure by the following example.

In order to force $\pi = 0$ in (i), $c(\text{Na})$ in phase B must be reduced by 11 mmol/l, or by less than 10% of the previously determined concentration of 128 mmol/l.

Introducing this change, there is no longer a state of equilibrium across the membrane, and other transport processes will take place. The treatment of such non-equilibrium states is postponed until Chapter 7.

Example 6.7

The electrical potential difference between the interior of a cell and its surrounding liquid is determined as 90 mV, with the cell interior negative, $\Delta\mathscr{E} = \mathscr{E}_B - \mathscr{E}_A = -90$ mV. The concentrations of existing ions Na^+, K^+ and Cl^- in the cell interior are 12, 139 and 4 mmol/l, respectively, and in the surrounding liquid, 145, 4 and 116 mmol/l respectively. The activity coefficients for these ions are assumed to be the same in both phases and the temperature is 37 °C. We wish to estimate which of the three ions are close to equilibrium.

Using (a) in Example 6.6 for each of the three ions we can calculate the membrane potential that would exist, given the concentrations, if there was equilibrium:

$$\Delta\mathscr{E}\,(Na) = [61.5/(+1)]\log(Na_A/Na_B) = +66.5\,\text{mV}, \qquad (a)$$

$$\Delta\mathscr{E}\,(K) = [61.5/(+1)]\log(K_A/K_B) = -94.7\,\text{mV}, \qquad (b)$$

$$\Delta\mathscr{E}\,(Cl) = [61.5/(-1)]\log(Cl_A/Cl_B) = -90.0\,\text{mV}. \qquad (c)$$

Here, the numerical value $\hat{R}T/\mathscr{F} = 8.314 \times 310 \times 100 \times \ln(10)/96\,500 = 61.5$ mV has been used. Cl^- is therefore at equilibrium, K^+ is close to equilibrium, while Na^+ is far from equilibrium. The explanation is that Na^+ is transported actively from the interior of the cell to the surrounding liquid by means of a chemical pump (see Example 7.5).

Example 6.8

A membrane, permeable to hydrogen ions but not to chloride ions, separates two aqueous solutions of HCl, denoted A and B. The two solutions have the same pressure (1 atm) and temperature (25 °C), but different HCl concentrations, being 10 mmol/l and 1 mmol/l, respectively for A and B. We wish to calculate the electrical potential difference between the two solutions at equilibrium, as well as the number of hydrogen ions that have moved across the membrane during the time taken to reach equilibrium.

For the only permeating component H^+, the potential difference is calculated using (a) in Example 6.7:

$$\mathscr{E}_B - \mathscr{E}_A = [59/(+1)]\log(10/1) = 59\,\text{mV}. \tag{a}$$

Biological membranes have an electrical capacity C_m of about $1\,\mu\text{F/cm}^2$. A potential difference of $59\,\text{mV} = 0.059\,\text{V}$ will therefore induce a charge density of

$$q = C_m\,V = 10^{-6} \times 0.059 = 5.92 \times 10^{-8}\,\text{coulomb}. \tag{b}$$

One mole of protons carries a charge of $96\,485 \approx 10^5$ coulomb. The number of protons that have moved is therefore $5.9 \times 10^{-13}\,\text{mol/}$ cm^2, or $5.9 \times 10^{-13} \times 6.023 \times 10^{23} \approx 3.6 \times 10^{11}$ protons/cm^2.

6.5 Chemical reaction equilibrium. The law of mass action

Among the alternative conditions for equilibrium that are derived in Section 6.1.2 (cp. Table 6.1), the condition $dA_{T,V} = 0$ can be used to determine the equilibrium composition of a reactive mixture enclosed in a vessel at fixed temperature and volume. Condition (15), $dG_{T,p} = 0$, determines the composition of a reactive mixture at equilibrium at fixed temperature and pressure, a condition of special interest for biological systems.

Substitution of (3.89) into (15) gives, for $dT = 0$ and $dp = 0$,

$$dG_{T,p} = \Sigma\;\hat{\mu}_j\;dN_j = 0. \tag{52}$$

Here, only the chemical potential is included since the electrical potential energy vanishes for a homogeneous mixture which fulfils the condition of electroneutrality.

For a homogeneous non-reactive mixture, all mole numbers are constant, $dN_j = 0$, so that (52) is valid for arbitrary values of the chemical potentials.

For a reactive mixture, on the other hand, the mole numbers can change in the direction of products or reactants when the reaction proceeds towards equilibrium. The summation in (52) includes all the components in the mixture, $j = 1,2,\ldots,k$, that enter into the chemical reaction and there will be both reactants and products at equilibrium. In contrast to (3.77), the reaction can be conveniently written in the form

$$0 \leftrightarrows \Sigma \, \nu_j \, B_j. \quad \nu_j \begin{cases} < & 0 \text{ reactant} \\ = & 0 \text{ non-participant} \\ > & 0 \text{ product} \end{cases} \qquad (53)$$

Since the mass of the systems is constant, the change in mole numbers of the various components is not arbitrary but is determined by the stoichiometric coefficients ν_j in (53), which expresses conservation of atom numbers.

The equilibrium composition N_j can be expressed in terms of an initial composition N_{j0} and a single quantity, the *extent* ε, as

$$N_j = N_{j0} + \nu_j \, \varepsilon, \qquad (54)$$

or, after division by the total number of moles, $\Sigma \, N_j$, by mole fractions,

$$\chi_j = (N_{j0} + \nu_j \, \varepsilon)/\Sigma \, N_j, \qquad (55)$$

or, for dilute solutions, after division by the volume, by concentrations,

$$c_j = c_{j0} + \nu_j(\varepsilon/\mathcal{V}), \qquad (56)$$

where $\varepsilon/\mathcal{V}$ can be considered another extent. In the above expressions the stoichiometric coefficients follow the sign convention of (53).

The differential of (54):

$$dN_j = \nu_j \, d\varepsilon, \qquad (57)$$

is now inserted into (52) and the change $d\varepsilon$, which is arbitrary at equilibrium, is dropped, leaving the *condition for chemical equilibrium* in a reactive mixture with a single reaction:

$$\Sigma \nu_j \, \hat{\mu}_j(T,p,\chi_j) = 0 \quad \text{(reaction equilibrium)}. \qquad (58)$$

When the temperature, pressure and the initial composition (or the ratios of the atom numbers for the whole mixture) are specified, (58) determines the specific composition of the mixture at equilibrium, expressed by the mole fractions χ_j, $j = 1,2,\ldots,k-1$. When several reactions proceed simultaneously, equation (58) applies to each reaction. The final form of (58) depends on the relevant expression for the chemical potential of components, and is illustrated below.

In general, substituting (3.121) into (58) and rearranging, we obtain *the law of mass action* of the form

$$\Pi \, (a_j/a_j^0)^{\nu_j} = K \text{ (in general).} \tag{59}$$

Here, K denotes the equilibrium constant

$$K = \exp \, (-\Delta \hat{g}_r^0/\hat{R}T), \tag{60}$$

and the standard reaction Gibbs energy is

$$\Delta \hat{g}_r^0 = \Sigma \, \nu_j \, \hat{\mu}_j^0, \tag{61}$$

as calculated on the basis of the chemical potentials $\hat{\mu}_j^0 = \hat{g}_j(T, p_0)$ at the standard state, which defines the standard activities a_j^0.

According to Table 3.3, we obtain a first special case for the reactive *ideal gas mixture*, employing

$$a_j/a_j^0 = p_j/p^0 = (p_j/p)(p/p_0) = \chi_j \, p/p_0. \tag{62}$$

Substituting this into (59) and rearranging gives the law of mass action expressed in terms of mole fractions:

$$\Pi \chi_j^{\nu_j} = (p/p_0)^{-\Sigma \, \nu_j} \, K(T) \text{ (gas).} \tag{63}$$

Since $\hat{\mu}_j^0 = \hat{g}(T, p_0)$ in (61) with $p_0 = 1 \, \text{atm}$, also cp. (3.120), (60) shows that $K(T)$ can be calculated from the state properties of the pure components and that it depends on the absolute temperature only. Therefore, $K(T)$ determines the effect of temperature on the equilibrium, while the effect of pressure has been isolated in the first term on the right-hand side of (63). In biological systems, usually $p \approx p_0$ so the pressure effect is not important. Introducing $\chi_j = p_j/p$ and $p \approx p_0$, (63) can be expressed in terms of the partial pressure p_j:

$$\Pi p_j^{\nu_j} = p_0^{-\Sigma \, \nu_j} \, K(T). \tag{64}$$

According to Table 3.3, we obtain a second special case with $a_j/a_j^0 = \chi_j$ for the *ideal liquid*:

$$\Pi \chi_j^{\nu_j} = K(T) \text{ (liquid)} \tag{65}$$

which is (63) without pressure dependence.

As a final case, $a_j/a_j^0 = c_j/1.0$ substituted into (59) gives

$$\Pi c_j^{\nu_j} = K(T) \text{ (aq. solution),} \tag{66}$$

where $\hat{\mu}_j^0$ in (61) is valid for the standard state $c_j^0 = 1.0 \, \text{mole per litre at}$

temperature T, and $K(T)$ has the dimension mol/l raised to the power $\Sigma \nu_j$.

Alternative expressions arise by using activity coefficients in (59), according to (3.124) or (3.125). These resulting expressions reduce to (65) and (66), respectively, for ideal cases.

The quantity on the left-hand side of (59) and (63)–(66) is denoted as the *activity ratio* of the reaction. For a non-equilibrium state, this ratio differs from the right-hand side of the equation in question. If this ratio is larger than the right-hand side, the reaction must proceed towards the reactants in order to establish equilibrium, cp. the algebraic sign of ν_j in (53). On the other hand, if the ratio is smaller than the right-hand side, the reaction proceeds towards the products. Such displacements towards equilibrium follow the principle of Le Chatelier.

If calculations show that $K(T)$ is a very large number, the equilibrium state is more or less totally displaced in favour of the products. On the other hand, if $K(T)$ is very small, the state is displaced in favour of the reactants. It should be noted that if products and reactants are interchanged in the statement of the reaction equation ((53)), all stoichiometric coefficients change sign and, according to (60), $K(T)$ for this reverse reaction is the reciprocal of $K(T)$ for the original reaction.

The temperature dependence of the equilibrium constant is given by (60), where the standard reaction Gibbs function for the ideal case can be calculated by using (5.30), and (d) in Example 4.15 and (d) in Example 5.7:

$$\Delta \hat{g}_r^0(T,p_0) = \Delta \hat{g}_r^0(T_0,p_0) + [(T - T_0) - \ln (T/T_0)] \Sigma \nu_j \, \hat{c}_{pj}. \qquad (67)$$

However, it is often more convenient to use the general *van't Hoff equation*,

$$d \ln [K(T)]/dT = \Delta \hat{h}_r^0(T)/[\hat{R}T^2], \qquad (68)$$

which is derived for constant pressure by differentiation of (60) and use of (5.30) and (35) for each component as pure matter. For a small change in temperature, changes in reaction enthalpies are small (cp. Example 4.15) so that (68) can be integrated to

$$\ln [K(T)/K(T_0)] \approx [(T - T_0) \Delta \hat{h}_r^0(T_0)]/(\hat{R}TT_0). \qquad (69)$$

Inserting (60) into (69), we obtain an alternative to (67):

$$\Delta \hat{g}_r^0(T,p_0) \approx (T/T_0) \Delta \hat{g}_r^0(T_0,p_0) - (T/T_0 - 1) \Delta \hat{h}_r^0(T_0). \qquad (70)$$

Note that the van't Hoff relation ((68)) is of considerable importance in quite a different context. Thus, if the chemical composition of a given reactive mixture is determined at equilibrium at a number of different temperatures T, the value of $K(T)$ can be calculated from (59), or (63)–(66), and then differentiated according to (68) to determine the reaction enthalpy. Since the reaction Gibbs function can be determined simultaneously by (60), the reaction entropy is obtained.

As mentioned already, Le Chatelier's principle concerning equilibrium displacement can be illustrated by (59). The effect of a temperature change is obtained from (68). This equation shows, for example, that a negative reaction enthalpy, corresponding to an exothermic reaction, implies a decreasing $K(T)$ with increasing T so, imposing a temperature increase would shift the composition in favour of the reactants. This response opposes the imposed change. For an endothermic reaction, the reverse response would result. In other words, the drive towards equilibrium would have a direction that opposes the externally imposed change. Similarly, increasing the concentration of a component would cause a shift in equilibrium so as to diminish the concentration of the component.

Example 6.9

Calculate the equilibrium constant $K(T)$ at $T = 25\,°C = 298\,K$ for the gaseous reaction

$$N_2O_4 \rightleftarrows 2NO_2, \tag{a}$$

using $\hat{g}_1 = \hat{g}(N_2O_4) = 98\,kJ/mol$ and $\hat{g}_2 = \hat{g}(NO_2) = 52\,kJ/mol$ at $25\,°C$.

The reaction Gibbs function is calculated from (58):

$$\Delta\hat{g}_r = -1 \times 98 + 2 \times 52 = 6\,kJ/mol\ N_2O_4, \tag{b}$$

and (60) gives

$$K(298\,K) = \exp[-6000/[(8.314 \times 298)]] = 0.0888. \tag{c}$$

The equilibrium constant for the reverse of reaction (a) is the reciprocal of (c), $1/0.0888 = 11.36$.

Example 6.10

Pure N_2O_4 is brought to the state $T = 25\,°C$ and $p = 0.1\,atm$, after which equilibrium is established according to the reaction

$$N_2O_4 \rightleftarrows 2\,NO_2. \tag{a}$$

Calculate the resulting composition when the equilibrium constant, according to Example 6.9, is $K(298) = 0.0888$. The stoichiometric coefficients are $\nu_1 = \nu(N_2O_4) = -1$ and $\nu_2 = \nu(NO_2) = +2$, and $\Sigma\nu_j = -1 + 2 = +1$, so that (59) takes the form

$$\chi_2^2/\chi_1 = (p/p_0)^{-1} K(T). \tag{b}$$

The atom number ratio N/O is given as $\frac{2}{4}$ and the initial mole numbers can be given as $N_{10} = 1$ and $N_{20} = 0$. According to (54), the final state can be expressed as

$$N_1 = 1 - \varepsilon; \quad N_2 = 2\,\varepsilon, \tag{c}$$

or, according to (55), using $\Sigma\, N_j = 1 - \varepsilon + 2\,\varepsilon = 1 + \varepsilon$, as

$$\chi_1 = (1 - \varepsilon)/(1 + \varepsilon); \quad \chi_2 = 2\,\varepsilon/(1 + \varepsilon), \tag{d}$$

which, substituted into (b), gives

$$4\varepsilon^2/(1 - \varepsilon^2) = (p/p_0)^{-1} K(T). \tag{e}$$

With the right-hand side equal to $(1.0/0.1) \times 0.0888 = 0.888$, ε is calculated from (e) and the mole fractions from (d), as

$$\varepsilon = 0.426; \quad \chi_1 = 0.402; \quad \chi_2 = 0.598.$$

Inspection of (b) reveals that an increase in pressure will decrease the right-hand side and, accordingly, decrease χ_2, which illustrates Le Chatelier's principle. An equilibrium will be displaced in the direction that decreases the imposed change; in the case considered here, the number of moles will decrease by a displacement in favour of the reactant. Since $\Delta\hat{g}_r$ is positive, it is also evident that the value of $K(T)$ decreases with increasing temperature (see Example 6.9), leading to a decrease in ν_2. This displacement in favour of reactants is in agreement with the fact that the reaction is exothermic. An equilibrium is displaced in a direction (energy absorption) that diminishes the imposed change (increase in temperature).

Example 6.11

The formation of adenosinediphosphate ADP and fructosediphosphate FDP from adenosinetriphosphate ATP and fructose-6-phosphate is described by the reaction

$$ATP + F6P \rightleftarrows ADP + FDP, \tag{a}$$

whose standard reaction Gibbs energy is $\Delta \hat{g}_r^0 = -14.9\,\text{kJ/mol}$ at 37 °C.

Calculate the concentration of F6P at equilibrium in an experiment where ATP aned F6P are added to a solution to a concentration of $c_{10} = 2.0 \times 10^{-3}\,\text{mol/l}$ and $c_{20} = 4.5 \times 10^{-3}\,\text{mol/l}$ respectively.

Denoting components $j = 1,\ldots,4$ numbered according to (a), from left to right, the initial state for components 1 and 2 is given above, while $c_{30} = 0$ and $c_{40} = 0$. Since the solution is highly diluted, we use the extent $f = \varepsilon/\mathcal{V}$ according to (56) and obtain the equilibrium composition

$$c_1 = c_{10} - f; \quad c_2 = c_{20} - f; \quad c_3 = f; \quad c_4 = f, \qquad \text{(b)}$$

which inserted into (66) gives

$$f^2/[(c_{10}-f)(c_{20}-f)] = \exp(-\Delta\hat{g}_r^0/\hat{R}T). \qquad \text{(c)}$$

The right-hand side of (c) is $K(310\,\text{K}) = \exp[-14\,900/(8.314 \times 310)] = 324$. From (c) we then obtain $f = 2.5 \times 10^{-3}\,\text{mol/l}$, so that the concentration of F6P is obtained by (b) as

$$c_2 = c_{20} - f = (4.5 - 2.5) \times 10^{-3} = 2.0 \times 10^{-3}\,\text{mol/l}. \qquad \text{(d)}$$

Example 6.12

Calculate the ion product of water at 25 °C, i.e. the product of the concentrations of H^+ and OH^-, or $c_2 c_3$.

Assuming dissociation equilibrium according to the reaction

$$H_2O\ (1) \rightleftharpoons H^+\ (aq) + OH^-\ (aq), \qquad \text{(a)}$$

and inserting concentrations for ions and mole fraction for water, into (59), cp. Section 3.10.5, gives

$$c_2 c_3/\chi_1 = K(T), \qquad \text{(b)}$$

where the components are numbered in the sequence given by (a) from left to right. The initial state is taken to be pure water so that $c_{10} = 1000/18.02 = 55.5\,\text{mol/l}$. Inserting the extent f from (56), the composition at equilibrium, expressed in mole fraction and concentration, is

$$\chi_1 = 1 - f/55.5; \quad c_2 = f; \quad c_3 = f. \qquad \text{(c)}$$

The equilibrium constant $K(T)$ in (b) is calculated from (60), with the reaction Gibbs energy for (a) from (61). At the standard state (25 °C, 1 atm and 1 mol/l) we obtain by using Appendix C.1,

$$\Delta \hat{g}_r^0 = - (-237) + 0 + (-157) = 80 \, \text{kJ/mol}, \tag{d}$$

and

$$K(298 \, \text{K}) = \exp[-80\,000/(8.314 \times 298)] = 0.948 \times 10^{-14}. \tag{e}$$

It can be assumed that $f \ll 55.5$, so that $\chi_1 \approx 1$ in (c), and (b) gives

$$c_2 c_3 = 1 \times K(T) = 0.948 \times 10^{-14} \, (\text{mol/l})^2. \tag{f}$$

From this, it follows that $c_2 = c_3 = 0.97 \times 10^{-7} \approx 10^{-7} \, \text{mol/l}$, which is pH = 7 for water, cp.(3.127).

To show the effect of temperature, we also determine the ion product at 37 °C, $T = 310 \, K$ and $p_0 = 1$ atm. Inserting values from Appendix C.1, the standard reaction enthalpy is calculated as

$$\Delta \hat{h}_r^0(T_0) = - (-286) + 0 + (-230) = 56 \, \text{kJ/mol}, \tag{g}$$

and (70) gives

$$\Delta \hat{g}_r^0(T, p_0 20) = (310/298) \times 80 - (310/289 - 1) \times 56 = 81 \, \text{kJ/mol}, \tag{h}$$

Accordingly, (e) and (f) are changed to give

$$K(310 \, \text{K}) = c_2 c_3 = 2.27 \times 10^{-14}, \tag{i}$$

corresponding to pH = 6.8.

7

Thermodynamic non-equilibrium

It is a characteristic feature of all forms of life that processes take place continuously. These are both local, i.e. chemical reactions, and global, i.e. transport processes associated with transfer of energy, matter and electrical charge in a given direction. The two types of processes often take place in structures that are semi-solid or solid and they are usually closely coupled. Here, however, we only treat each type of process separately. Further remarks on simultaneous local and global processes and structures are given in Section 8.6.

The occurrence of any process is a manifestation of some departure from thermodynamic equilibrium since equilibrium is defined as a state at which no process can take place (Chapter 6). Furthermore, the occurrence of processes implies irreversibilities (Chapter 5), since we are dealing with spontaneous events.

Taking the production of entropy as a measure of irreversibility, we can identify fluxes and associated potentials, e.g. a heat flux and a difference or gradient in temperature. According to the thermodynamics of irreversible processes, fluxes are postulated as being linear functions of related potentials. From this follows phenomenological constitutive relations for fluxes which formally fit well with empirically determined constitutive relations, e.g. Fourier's law of heat conduction. Transport properties in these relations, e.g. thermal conductivity, are constitutive, depend on the matter in question, and must be determined by experiment, possibly supplemented by models and theories of molecular behaviour.

Transport processes across membranes are of particular importance for living systems and are treated in a separate section. In addition, we analyse in detail a simple model of transport processes of ions in a typical living cell.

7.1   Local non-equilibrium. Chemical reaction

The condition for equilibrium in a simple homogeneous system (Section 6.1) can alternatively be expressed, depending on constraints, as a maximum of entropy or a minimum of the internal energy, enthalpy, Helmholtz function or Gibbs function, respectively (Table 6.1). These results follow from the first and the second laws, since the state of the homogeneous system will approach the equilibrium states mentioned as long as there are spontaneous processes in the system.

As a concrete example, let a homogeneous mixture of chemically reactive species suddenly be established in a closed system (Figure 7.1). The so far unspecified spontaneous processes would be one or several chemical reactions whose rates would depend on, e.g., the composition of the mixture, presence of catalysts, the temperature and pressure. For a single reaction ((3.77)) with a molar reaction rate ξ from (3.78), we have conservation of mass (3.73) with $r = 1$. In addition, we have the first law ((4.1)), without changes in kinetic and mechanical potential energy, and the balance of entropy ((5.6)), i.e.

$$\mathrm{d}N_i/\mathrm{d}t = \nu_i\,\xi; \quad \mathrm{d}\tilde{U}/\mathrm{d}t = \dot{Q} + \dot{W}; \quad \mathrm{d}S/\mathrm{d}t = \dot{Q}/T + \dot{\sigma}. \tag{1}$$

Finally, we assume the Gibbs relation ((3.83)) to be valid:

$$\mathrm{d}\tilde{U}/\mathrm{d}t = T\,\mathrm{d}S/\mathrm{d}t - p\,\mathrm{d}\mathcal{V}/\mathrm{d}t + \Sigma\,\tilde{\mu}_j\,\mathrm{d}N_j/\mathrm{d}t, \tag{2}$$

where the contribution from electrical potential energy, cp.(3.137), is included on account of the treatment in the next section, though such contributions are without importance for simple chemical reactions.

According to the concept of state ((2.3)), the form of (2) is valid for a homogeneous state of equilibrium, cp. Section 3.10. When the rate of

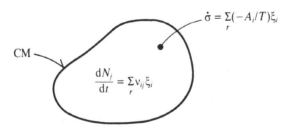

Fig 7.1 Local (homogeneous) non-equilibrium in control mass with mixture of components $j = 1, 2, \ldots, k$ in which reactions $i = 1, 2, \ldots, r$ occur with reaction rates ξ_i and affinities A_i.

chemical reaction is slow, it is reasonable to think of the process as proceeding through states that are in *thermomechanical equilibrium* at all times in spite of the chemical non-equilibrium. This idea can be readily understood for a gaseous mixture when non-reactive collisions are far more frequent than reactive collisions. Here, the molecular distribution functions that describe the thermochemical quantities temperature and pressure are effectively the same as those at equilibrium with the given composition. Such conditions are also valid with good approximation for real reaction rates in both gases and in liquid solutions. This situation makes it possible to define and calculate all the necessary properties of state, including the electrochemical potential $\bar{\mu}_j(T,p,\chi_j)$, even though the composition of the mixture χ_j does not fulfil the condition of chemical equilibrium.

Substitution of (1) into (2), and assuming that work is associated only with reversible change in volume, leads to the following expression for entropy production:

$$\dot{\sigma} = (-A/T)\,\xi \geqslant 0, \tag{3}$$

where A denotes the chemical *affinity* of the reaction at its instantaneous non-equilibrium state,

$$A = \Sigma\, \nu_j\hat{\mu}_j. \tag{4}$$

It is important to realize the difference between the chemical affinity (4) and the reaction Gibbs function given in (5.30) as $\Delta\hat{g}_r = \Sigma\nu_j\hat{\mu}_j$. While the stoichiometric coefficients ν_j are the same in the two expressions, the chemical potentials are different.

In the expression for $\Delta\hat{g}_r$, the chemical potentials refer to reactants at their state prior to reaction and to products at their state after completion of the reaction respectively. Thus $\Delta\hat{g}_r$ is the change in Gibbs function per mole fuel, say for the complete net reaction considered, for example (4.46) for the combustion of glucose,

$$\Delta\hat{g}_r = (G_{\text{products}} - G_{\text{reactants}})/N_{\text{fuel}}.$$

This reaction, say at constant T,p, could take place in a steady flow process from inlet to outlet of a control volume, or it could proceed as an unsteady process in a control mass from an initial state of reactants to a final state of the products of the net reaction. In (4), on the other hand, the chemical potentials of reactants and products are their instantaneous values during the course of the reaction. This may be clarified by introducing the extent ε of the reaction given in (6.54), noting also that the instantaneous reaction rate is simply $\xi = d\varepsilon/dt$, according to (3.78) and (6.57). Also, according to (3.87) we have $G = \Sigma\hat{\mu}_j N_j$, and (3.89), after substitution of (6.57) and (4), shows that

$dG_{T,p} = \Sigma \, \hat{\mu}_j dN_j = (\Sigma \nu_j \hat{\mu}_j) d\varepsilon = A \, d\varepsilon$, implying that $A = (\partial G/\partial \varepsilon)_{T,p}$. Thus, as the Gibbs function G of a reactive mixture changes as the reaction proceeds at constant T,p in a steady control volume process, or in an unsteady control mass process, the chemical affinity, which equals the slope of G versus ε, changes (see also Welch, 1985).

The values $\varepsilon = 0$ and $\varepsilon = 1$ correspond to pure reactants and pure products respectively. An intermediate value $0 < \varepsilon < 1$ corresponds to a state of partial or incomplete reaction. In some cases such a state may be that of chemical equilibrium, in which case the state will be characterized by $A = 0$, cp. (6.52).

For several simultaneous chemical reactions, $i = 1,2,\ldots,r$, the right-hand side of the molar balance in (1) will involve a summation, (3.80), so that (3) now involves a sum of the products of reaction rates and affinities:

$$\dot{\sigma} = \Sigma_r \, (-A_i/T) \, \xi_i \geqslant 0, \tag{5}$$

where the affinity for each reaction is defined as in (4).

Result (3) can be interpreted as follows. At equilibrium, no process takes place, $\xi = 0$, the entropy production is zero, and $A = 0$, in accordance with (6.58), which is the condition for chemical equilibrium. The Gibbs function of the system then attains its minimal value at the given values of T and p. When the reaction proceeds, the second law prescribes that $\dot{\sigma} \geqslant 0$, so that if $A \leqslant 0$, we must have $\xi \geqslant 0$, in accordance with (3), which shows that the reaction proceeds with formation of products. This also agrees with the general condition ((6.14)) that the Gibbs function of the system must decrease during processes towards equilibrium at fixed values of T and p. The same conclusion is reached for the case where $A > 0$, where (3) requires that $\xi \leqslant 0$, showing that the reaction proceeds with formation of reactants.

The rate of the reaction is determined empirically. Theoretical considerations indicate that the rate depends on concentrations and temperature in a way that is expressed by the Arrhenius equation:

$$\xi = D \exp \, (-E_a/RT), \tag{6}$$

where D depends, among other things, on the presence of catalysts and the concentrations of the components in the mixture, and E_a is the *energy of activation* for the critical, i.e. rate limiting, step in the reaction. The mechanism for the effect of concentrations is expressed in Guldberg and Waage's law of mass action, according to which the rate of the reaction is proportional to the 'active masses', i.e. mole fractions or concentrations of the components. The basic idea here is that the mole fractions determine the frequency of collisions between the components and thereby the probability that the reaction proceeds.

Affinity (4) can usually be calculated from thermodynamic data for the composition of the mixture, its temperature and pressure. The reaction rate, on the other hand, can only be calculated in a few cases from theories of reaction kinetics, a complex and expanding field of enquiry (for an introduction, see, e.g., Prigogine and Defay, 1954; Rock and Gerholdt, 1974).

Useful expressions for the affinity follow the transformation of (6.58) into (6.59), or, by use of concentrations, into (6.66). Thus, using (3.121) and $a_j = c_j$ in accordance with Table 3.1, (4) can be written as

$$\prod_{1}^{k} c_j^{\nu_j} = \exp\left[(A - \Delta \hat{g}_r^0)/\hat{R}T\right], \tag{7}$$

which reduces to (6.66) at equilibrium when $A = 0$. Note that the non-equilibrium concentrations c_j in (7) are different from the equilibrium concentrations c_j in (6.66) and that they approach the latter as equilibrium is approached.

We note, furthermore, that the entropy production (3) can be viewed as a product of a *driving force*, affinity or difference in potential $(-A/T)$, which is an intensive property, and a *flux*, the reaction rate, which is an extensive property. Such considerations form the basis for the phenomenological derivation of constitutive relations for fluxes within the framework of irreversible thermodynamics (Section 7.3).

In real systems, e.g. living cells, the resulting reaction rates are also determined by the rate of transport of matter by way of bulk flow and diffusion to and from the localities of chemical reactions. Here, one can distinguish between two extreme cases: that where the resulting rate is reaction limited and that where it is transport limited. In the first case, the intrinsic chemical reaction rate is small and determines the resulting rate. This case corresponds to the conditions of a homogeneous reaction where the relatively fast transport processes will suppress any concentration gradients and ensure a spatially homogeneous state. In the second case, the intrinsic reaction rate is large and the rate of transport of reactants and products to and from the locality of the chemical reaction will determine the resulting rate. In living systems, both extreme cases occur and spatial non-equilibria form an important part of the treatment that follows.

7.2   Global non-equilibrium. Spatial transport processes

The condition of equilibrium in a spatially non-homogeneous system is derived in Section 6.2 for the case of two subsystems, each one being homogeneous (Figure 6.1). When there is free exchange of energy and matter across the boundary that separates the two subsystems, result (6.22) can be expressed in terms of several separate conditions for equilibrium, that is, thermal, mechanical, chemical and electrical equilibrium. At equilibrium, the subsystems will have equal values of temperature, pressure, and electrochemical potentials respectively. As equilibrium is approached, (6.22) shows this to occur from a state of spatial non-equilibrium associated with spatially non-uniform distributions of thermal, mechanical, electrical and chemical potentials, and with transport processes of heat, volume, electrical charge and matter.

Spatial non-equilibria in real systems can often be represented by spatial discontinuities in potentials across a barrier (e.g. a cell membrane) which separates two systems, each of which is homogeneous and at equilibrium. This approximation holds good if the major resistance to transport is located at the barrier.

In other cases, spatial non-equilibria are represented by spatially continuous distributions of potentials, such as gradients of temperature that give rise to heat conduction.

The two types, discrete and continuous spatial non-equilibrium, are treated separately in the following sections for the cases summarized in Figure 7.2.

7.2.1   Discrete spatial non-equilibria

Let us now repeat the derivation of (6.22) for unsteady processes in the isolated control mass C shown in Figure 6.1. For each of the systems A and B, we use (1) and (2), and for system C, we employ (6.16), (6.17), and the entropy balance $dS_C/dt = \dot{\sigma}_C$. This gives the total entropy production

$$\dot{\sigma}_C = \left(\frac{1}{T_A} - \frac{1}{T_B} \right) \frac{d\tilde{U}_A}{dt} + \left(\frac{p_A}{T_A} - \frac{p_B}{T_B} \right) \frac{d\tilde{V}_A}{dt}$$

$$- \sum \left(\frac{\tilde{\mu}_{jA}}{T_A} - \frac{\tilde{\mu}_{jB}}{T_B} \right) \frac{dN_{jA}}{dt} \geq 0, \tag{8}$$

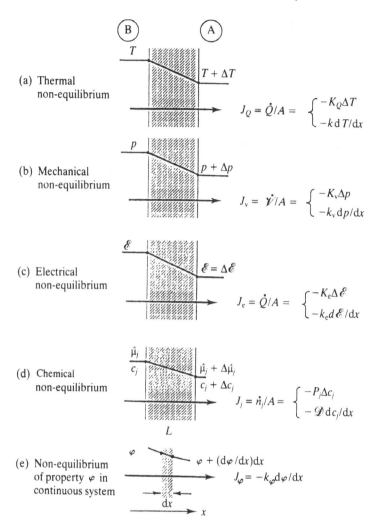

Fig 7.2 A schematic illustration of individual non-equilibria in discrete systems (a)–(d) and continuous systems (e) with transport of heat, volume, electrical charge and matter. Fluxes J are transported quantities per unit of time and area.

which must take place in the barrier since subsystems A and B are assumed to be homogeneous and at equilibrium.

At non-equilibrium we first show that (8) can describe four different transport processes, each of which occurs separately for certain specific internal restrictions on the character of the barrier (cp. Section 6.2). Each of the terms in (8) is a product of an intensive quantity, the driving force for transport, and an extensive quantity, the flux of the transported quantity. Fluxes for transport are considered to be positive from B to A and the driving forces are differences in potentials between A and B, see Figure 7.2.

Figure 7.2(a) shows the *thermal non-equilibrium* $T_A - T_B = \Delta T$ across a non-movable, impermeable, but heat conducting, barrier. Since $\dot{W} = 0$ in (1), and $dU_A/dt = \dot{Q}$, the flux is simply the heat flux which, per unit of area, can be expressed as

$$J_Q = \dot{Q}/A = -K_Q \Delta T, \tag{9}$$

where K_Q is a coefficient of heat conductance for the barrier. The empirical form ((9)) is used, for example, for convective heat transfer, cp. (3.27), with $K_Q = \bar{h}$. The entropy production per unit of area is calculated from the first term of (8):

$$\dot{\sigma}_Q/A = [K_Q/(T_A T_B)] (\Delta T)^2, \tag{10}$$

which shows that K_Q must be positive.

We now consider the mechanical non-equilibrium $p_A - p_B = \Delta p$ across a movable, impermeable and non-conducting barrier. Since system C in Figure 6.1 is isolated from the surroundings, its volume is constant. Relaxation of the pressure difference results in a displacement of the barrier, and hence compression or expansion of subsystems A and B, as described in Example 5.4. This form of mechanical non-equilibrium is not of particular interest in biological systems. However, a very different consideration, where the change in volume is treated as a flow of volume across a control volume, leads to the idea that $(p_A - p_B)\ \dot{V}_A$ can be considered as the difference in the flow work, cp. (4.6), associated with the outflow and inflow, respectively, across a permeable barrier. The pressure difference is now due to viscous fluid friction and we have a transport process of matter.

Figure 7.2(*b*) shows this case of *mechanical non-equilibrium* for an isothermal, one-component system with flow through a non-movable, permeable barrier. For an incompressible viscous fluid, the volume flow per unit of area, the permeate flux or mean velocity V, can be expressed, as for laminar flow in channels (cp. (4.28) and (4.31)), as

$$J_v = (1/A)\ d\dot{V}_A/dt = V = -P_v \Delta p, \tag{11}$$

where P_v is a *permeability coefficient* which is proportional to the porosity of the barrier and inversely proportional to the viscosity of the fluid. The entropy production per unit of area can then be calculated from the penultimate term in (8), using (11):

$$\dot{\sigma}_v/A = -J_v\,\Delta p/T = (P_v/T)\,(\Delta p)^2, \tag{12}$$

which shows that P_v is positive, as expected from (11).

Figure 7.2(c) shows the *electrical non-equilibrium* $\mathscr{E}_A - \mathscr{E}_B = \Delta\mathscr{E}$ in an isothermal system across a non-movable barrier which has an electrical conductance K_e and is permeable only to electrons. The flux is the electrical current per unit area, which satisfies Ohm's law:

$$J_e = -z_e\mathscr{F}\,(\mathrm{d}N_e/\mathrm{d}t)/A = -K_e\,\Delta\mathscr{E}, \tag{13}$$

where index e represents electrons. The entropy production is calculated from the last term in (8), with $\tilde{\mu}_A - \tilde{\mu}_B = z\mathscr{F}\,(\mathscr{E}_A - \mathscr{E}_B)$ according to (3.136) for constant T and p, and using (13):

$$\begin{aligned}\dot{\sigma}_e/A &= -(z_e\mathscr{F}/T)\,(\mathscr{E}_A - \mathscr{E}_B)\,(\mathrm{d}N_e/\mathrm{d}t)/A = -J_e\,\Delta\mathscr{E}/T \\ &= (K_e/T)\,(\Delta\mathscr{E})^2,\end{aligned} \tag{14}$$

which shows that K_e in (13) is positive.

Figure 7.2(d) shows *chemical non-equilibrium*, $\hat{\mu}_{jA} - \hat{\mu}_{jB} = \Delta\hat{\mu}_j$ or $c_{jA} - c_{jB} = \Delta c_j$, across a non-movable and semi-permeable barrier in an isothermal and isobaric system with the resulting transport of a single component. The flux, i.e. the flow of moles of components j from B to A per unit area, can be written as

$$J_j = (1/A)(\mathrm{d}N_{jA}/\mathrm{d}t) = -P_j\,\Delta c_j, \tag{15}$$

where P_j is a *permeability coefficient* for component j. The associated entropy production is calculated from the last term in (8). At constant temperature and pressure, the use of (3.121), for example for an aqueous solution (cp. Table 3.3), yields

$$\hat{\mu}_{jA} - \hat{\mu}_{jB} = \hat{R}T\ln\,(c_{jA}/c_{jB}) = \hat{R}T\ln\,(1 + \Delta c_j/c_{jB}), \tag{16}$$

which, for small differences in concentration, $c_{jA} \approx c_{jB} \approx c_j$, can be approximated as

$$\hat{\mu}_{jA} - \hat{\mu}_{jB} \approx \hat{R}T\,\Delta c_j/c_j. \tag{17}$$

Substitution of (17) and (15) into the last term of (8) gives

$$\dot{\sigma}_j/A = -J_j\,\hat{R}\,\Delta c_j/c_j = (P_j\,\hat{R}/c_j)\,(\Delta c_j)^2, \tag{18}$$

which shows that P_j in (15) is positive.

Before commenting on these results, we now re-examine the four transport processes for cases of continuous distributions of the driving potentials.

7.2.2 Continuous spatial non-equilibria

When the derivation of (8) is repeated for a differential control volume of thickness dx and area A, we again obtain a sum of contributions, each consisting of the product of a driving force and a flux. The potential $\varphi(x)$ is now assumed to be a continuous function of one spatial direction, x, and the change from φ at x to $\varphi + d\varphi$ at $x + dx$ can be expressed by a Taylor series to first order as $d\varphi = (d\varphi/dx)dx$, see Figure 7.2($e$). The entropy production in the differential volume $d\mathscr{V} = A\,dx$ can therefore be written generally as

$$d\dot{\sigma}_\varphi = \text{potential difference} \times \text{flux} = (d\varphi/dx)\,dx\,J_\varphi A, \qquad (19)$$

or, per unit volume, as

$$\dot{\sigma}/\mathscr{V} = \Sigma\,(d\varphi/dx)\,J_\varphi \geqslant 0, \qquad (20)$$

where the summation is over all φ.

For a heat flux, $J_Q = \dot{Q}/A$, and a flux of moles of component j, $J_j = \dot{n}_j/A$, i.e. contributions from the first and third term in (8), we can write (20) as

$$\dot{\sigma}/\mathscr{V} = \frac{d(1/T)}{dx}\,J_Q - \frac{d(\bar{\mu}_j/T)}{dx}J_j \geqslant 0. \qquad (21)$$

For *thermal non-equilibrium* considered separately, the heat flux in a homogeneous medium is expressed by Fourier's law (3.26):

$$J_Q = -k\,dT/dx, \qquad (22)$$

which, inserting (21), gives the entropy production

$$\dot{\sigma}/\mathscr{V} = (k/T^2)(dT/dx)^2 \geqslant 0, \qquad (23)$$

showing that the thermal conductivity k in (22) is positive. Contribution (23) is of quantitative importance in living systems only where there are large temperature gradients. An example is the blubber in seals, see Examples 3.9 and 5.2, where equation (23) is derived and given as (5.21). Unless stated otherwise, the following treatment deals with isothermal systems, $dT/dx = 0$.

In *electrochemical non-equilibria*, the molar flux of a component j is a result of a molecular transport, denoted diffusion, which can be ascribed

to the molecular thermal motion. The point of origin for a phenomenological description is the relation

$$flux = mobility \times concentration \times driving\ force, \tag{24}$$

which is Einstein's diffusion equation (1905):

$$J_j = - b_j\ c_j\ d\bar{\mu}_j/dx. \tag{25}$$

The driving force in an isothermal system is obtained from (21) as the
gradient in electrochemical potential. The flux is positive in the direction
of decreasing potential, according to experience (and the second law),
and is proportional to the local concentration. The constant of proportionality b_j is the *mobility* (the reciprocal value of the coefficient of
friction) of the transported molecules in the mixture. When (25) is
inserted into (21), the entropy production at T = constant becomes

$$\dot{\sigma}_j/\mathcal{V} = (b_j\ c_j/T)\ (d\mu_j/dx)^2, \tag{26}$$

which shows that b_j in (25) is positive.

The gradient of the electrochemical potential involves gradients in
pressure, concentration and electrical potential. This can be seen on
differentiation of (3.135), also using (3.136), (3.125), with $y_j = 1$ at
ideality, and $d\hat{\mu}_j^0 = -\hat{s}_j\ dT + \hat{v}_j\ dp = \hat{v}_j\ dp$ at T = constant:

$$\frac{d\bar{\mu}_j}{dx} = \hat{v}_j\ \frac{dp}{dx} + \frac{\hat{R}T}{c_j}\ \frac{dc_j}{dx} + z_j\mathcal{F}\frac{d\mathcal{E}}{dx}, \tag{27}$$

which, inserted into (25), gives the general result for the diffusion flux:

$$J_j = -b_j\ \hat{R}T\left(\frac{c_j\ \hat{v}_j}{\hat{R}T}\ \frac{dp}{dx} + \frac{dc_j}{dx} + \frac{c_j\ z_j\mathcal{F}}{\hat{R}T}\ \frac{d\mathcal{E}}{dx}\right). \tag{28}$$

In biological systems the term involving the pressure gradient is often
negligible. Then, diffusive transport of ions in an isothermal and isobaric
mixture is described by the *Nernst–Planck equation*

$$J_j = -b_j\ \hat{R}T\left(\frac{dc_j}{dx} + \frac{c_j\ z_j\mathcal{F}}{\hat{R}T}\ \frac{d\mathcal{E}}{dx}\right), \tag{29}$$

which shows that the diffusive flux of ions in a mixture depends on
gradients of concentrations and of electrical potential, contributions that
can work in the same or in opposite directions. If, for example, a mixture
of initially constant concentrations were subjected to a spatial gradient in
electrical potential, positive and negative ions would migrate in opposite

directions, and concentration gradients would arise (see Example 7.4). Conversely, if the mixture were subjected to concentration gradients, a spatial gradient in electrical potential would arise, provided that the mobilities of positive and negative ions were different (see Example 7.3). The gradient would attain a magnitude and an algebraic sign so as to make the fluxes of positive and negative ions equally large, and hence ensure electroneutrality locally. This kind of spatial difference in electrical potential, which can also be established across a membrane, is called a *diffusion potential*. It is an irreversible phenomenon, a result of a state of spatial non-equilibrium, in contrast to the membrane potentials, which are equilibrium potentials resulting from Donnan-equilibria, cp. Section 6.4.4.

For constant electrical potentials, $d\mathscr{E}/dx = 0$, or for non-electrolytes, (29) reduces to *Fick's law for diffusion*,

$$J_j = - D_j \, dc_j/dx, \qquad (30)$$

where

$$D_j = b_j \, \hat{R} T \qquad (31)$$

is the *diffusion coefficient* of the component. For gaseous mixtures, D_j is of the order of 10^{-6}–10^{-5} m^2/s. In liquids, D_j depends on (and decreases with increasing values of) the size of the molecules and the viscosity of the mixture. The value of D_j is also dependent on the shape of the molecules. Typical values for dilute aqueous solutions at 25 °C are, for water (self-diffusion) 2.4×10^{-9} m^2/s, for sucrose 5.2×10^{-10} m^2/s, and for haemoglobin 6×10^{-11} m^2/s ($\hat{M} = 64\ 000$ g/mol).

The term self-diffusion is used to describe the molecular transport in all spatial directions that always occurs in a homogeneous phase of matter, e.g. water. Because of the thermal fluctuations at the molecular level, a given molecule will move through the medium in random fashion. The same phenomenon will be observed for solutes in an aqueous solution of uniform concentrations.

Consider a one-dimensional system, e.g. a long tube in the direction of the x-axis that is filled with a mixture. At any cross-section along the tube, there will be a certain diffusive flux J_j^- in the negative x-direction and a corresponding diffusive flux J_j^+ in the positive x-direction. These fluxes are called *unidirectional fluxes*. The net flux in the x-direction, $J_j = J_j^+ - J_j^-$, is the diffusion flux described by (25) and it is zero for a pure substance or for a mixture of uniform electrochemical potential, where $J_j^+ = J_j^-$.

Consider the diffusive transport across the differential element dx in Figure 7.2(e). Here, J_φ^+ is proportional to the potential φ at x, while the flux J_φ^- is proportional to the potential $\varphi + (d\varphi/dx)dx$ at $x + dx$. It follows that the net flux $J = J_\varphi^+ - J_\varphi^-$ is proportional to $-(d\varphi/dx)$, as shown in the diagram. These considerations are valid for both continuous and discrete systems.

Diffusion coefficients for self-diffusion, or for diffusion of a given component in a mixture, can be determined by measuring the unidirectional flux of radioactively labelled, but otherwise identical, molecules (tracers). A flux of this kind will arise and be directed away from that part of the mixture that is suddenly labelled, while the oppositely directed unidirectional flux of labelled molecules will be zero initially (see Example 7.2).

A formal treatment of one-dimensional transport processes requires the use of mass conservation, including the contribution by diffusion. For component j in a mixture, (3.73) can be extended, in accordance with (3.6), to

$$dN_j/dt + \Sigma_{out}\, \dot{n}_j = \nu_j\, \xi - \Sigma_{out}\, A\, J_j. \tag{32}$$

where A denotes an area. For the special case of a steady process ($dN_j/dt = 0$) with no chemical reaction ($\xi = 0$) and flow of the mixture with constant velocity V in the x-direction ($\dot{n}_j = c_j\, V\, A$), equation (32) reduces to the contributions from flow and diffusion. The result, for a differential control volume element of thickness dx and area A (Figure 7.3), becomes, per unit area,

$$dw_j/dx = 0; \quad w_j = c_j\, V + J_j, \tag{33}$$

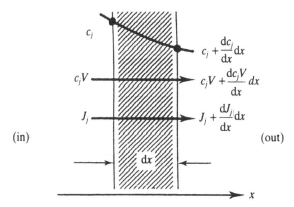

Fig 7.3 A differential control volume for stationary flow and diffusion in the x-direction.

where w_j represents the total transport (moles per unit area and time) in the x-direction.

We note that V is the velocity of the molar centre of the mixture, and (29)–(30) describe the diffusion relative to this centre. It follows that the sum of all diffusion fluxes for the mixtures is zero:

$$\Sigma J_j = 0, \tag{34}$$

so that (3.74) is also valid in the presence of diffusion processes. In a binary mixture, for example, $J_1 + J_2 = 0$, so that a diffusive flux of component 1 in one direction implies an equal flux of component 2 in the opposite direction.

Example 7.1

Figure 7.4 shows a diffusion cell of length $L = 0.005$ m situated between two large reservoirs, A and B, where the concentration of sucrose in water is kept constant, $C_{1A} = 0.1$ mol/l and $C_{1B} = 0.01$ mol/l. The mixture consists of non-electrolytes only. Calculate the diffusion flux J_1 of sucrose from A to B when $D_1 = 5 \times 10^{-10}$ m^2/s and the process is steady without flow in the cell. Calculate also the number of sucrose molecules that can be transported per second by diffusion across an area L^2 over a distance L, where $L = 2 \times 10^{-5}$ m is the typical size of cells. Integration of (30) for $J_1 = $ constant gives

$$J_1 = (D_1/L)\,(c_{1A} - c_{1B}) = (5 \times 10^{-10}/0.005)\,(0.1 - 0.01) \times 10^3$$
$$= 9 \times 10^{-6} \text{ mol/m}^2\text{-s.} \tag{a}$$

When L is of the order of cellular dimensions, the number of molecules transported per second is

$$J_1 = D_1\,L\,(c_{1A} - c_{1B})$$
$$= 5 \times 10^{-10} \times 2 \times 10^{-5} \times 0.09 \times 10^3 \times 6.023 \times 10^{23}$$
$$= 5.4 \times 10^{11} \text{ molecules/s.} \tag{b}$$

Example 7.2

The diffusion cell in Figure 7.4 now contains an aqueous solution of 100 mol/l NaCl. Then, 0.1 mmol radioactive Na with a specific activity of 10^8 units per mmol is added to reservoir A which has a volume of 1.0 l and is under continuous stirring. Continuous measurement of the radioactivity in reservoir A shows an initial

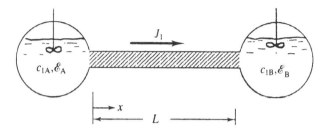

Fig 7.4 Diffusion cell with diffusion length L and constant concentrations and electrical potentials at $x = 0$ and L.

reduction in radioactivity of 14 units per minute. Determine the unidirectional flux of Na^+ from reservoir A, the diffusion coefficient D^+ and mobility b^+ of this ion when the temperature is 25 °C and the area of diffusion is $A = 100 \text{ mm}^2$. (In this and the following examples, '+' refers to cations '−' to anions and '1' to the salt.)

At any time, the measured specific activity is proportional to the concentration of radioactively labelled Na^+ in reservoir A, and at the time of addition it is $10^8 \times 0.1$ units. These measurements imply that, initially, the relative decrease in the concentration of labelled Na^+ in reservoir A is

$$(1/c^+_A) \, dc^+_A/dt = -14/10^7 = -1.4 \times 10^{-6} \text{ min}^{-1}, \tag{a}$$

and the concentration of labelled Na^+ is given as $c^+_A(0) = 0.1$ mmol/l.

Since there is no flow or chemical reaction, (32) gives for A as a control volume

$$dN^+_A/dt = -A \, J^+; \quad N^+_A = c^+_A \mathcal{V}, \tag{b}$$

from which, with (a),

$$
\begin{aligned}
J^+ &= -(\mathcal{V}/A) \, dc^+_A/dt = (10^{-3}/10^{-4}) \times 0.1 \times (1.4 \times 10^{-6}/60) \\
&= 2.33 \times 10^{-8} \text{ mol/m}^2\text{-s},
\end{aligned} \tag{c}
$$

which is the unidirectional flux of labelled Na^+. The unidirectional flux of Na^+ from the 100 mmol/l solution of NaCl in reservoir A is obtained by scaling the ratio between labelled and non-labelled Na^+:

$$J^+ = (100/0.01) \times 2.33 \times 10^{-8} = 2.33 \times 10^{-5} \text{ mol/m}^2\text{-s}. \tag{d}$$

An estimate of the diffusion coefficient is obtained by approximating (30) by

$$J^+ = -D^+(c_B^+ - c_B^+)/L = (D^+/L)\, c_A^+\,(0), \tag{e}$$

which implies a quasi-steady state during the measurement of the initial decrease in radioactivity and the assumption that the concentration of labelled Na^+ is negligible in reservoir B, which is also continuously stirred. From (c) and (e), using (a) and $L = 0.005\,m$ from Example 7.1, we obtain

$$\begin{aligned}
D^+ &= -(1/c_A^+)\,(dc_A^+/dt)\,(L\mathscr{V}/A) \\
&= (1.4 \times 10^{-6}/60)\,(0.005 \times 10^{-3}/10^{-4}) = 1.17 \times 10^{-9}\,m^2/s. \tag{f}
\end{aligned}$$

The mobility is then calculated from (31) as

$$\begin{aligned}
b^+ &= D^+/\hat{R}T = 1.17 \times 10^{-9}/(8.314 \times 298) \\
&= 4.7 \times 10^{-13}\,\text{mol-}m^2/\text{J-s}. \tag{g}
\end{aligned}$$

It should be noted that the expected value of $D^+ = 1.21 \times 10^{-9}\,m^2/s$, as calculated from (a) and (h) in Example 7.3 based on the value of $D_1 = 1.48 \times 10^{-9}\,m^2/s$ for NaCl at $25\,°C$, corresponds to an initial decrease in the radioactivity of 14.5 instead of 14 units per minute.

Example 7.3

The diffusion cell in Figure 7.4 now contains mixtures with fixed concentrations $c_{1A} = 100\,mmol/l$ and $c_{1B} = 10\,mmol/l$ of NaCl, respectively, in the two reservoirs. The mobilities of Na^+ and Cl^- are different and their ratio equals that of their transference numbers:

$$b^+/b^- = t^+/t^- = = 0.39/0.61\ (\text{NaCl}). \tag{a}$$

The *transference number* t of an ion equals the fraction of the total electrical current carried by the ion when the solution is subjected to an electrical potential gradient. It follows from (29) that this number is proportional to the mobility of the ion and its valency. Furthermore, for monovalent ions, cp. (44) in Section 7.2.3,

$$t^+ + t^- = 1. \tag{b}$$

We wish to calculate the diffusion potential of the cell at steady conditions, assuming a temperature of $25\,°C$ and activity coefficients that are equal in the two reservoirs.

As a consequence of the principle of electroneutrality, the local concentrations of Na^+ and Cl^- will be the same, say $c_1(x)$. Also, the fluxes of the two ions will be the same and directed from A to B. When this is used in (29), written for each of the ions Na^+ and Cl^- and after elimination of the fluxes, we obtain

$$\frac{d\mathscr{E}}{dx} = - \frac{b^+ - b^-}{b^+ z^+ - b^- z^-} \frac{\hat{R}T}{\mathscr{F}} \frac{1}{c_1} \frac{dc_1}{dx}. \tag{c}$$

This can be integrated over the cell to give the diffusion potential

$$\Delta\mathscr{E} = \mathscr{E}_A - \mathscr{E}_B = - \frac{b^+ - b^-}{b^+ z^+ - b^- z^-} \frac{\hat{R}T}{\mathscr{F}} \ln\,(c_{1A}/c_{1B}), \tag{d}$$

or, using (b) and $z^+ = z^- = 1$,

$$\Delta\mathscr{E} = (1 - 2t^+)\,(\hat{R}T/\mathscr{F})\ln\,(c_{1A}/c_{1B}), \tag{e}$$

which, with actual values, gives

$$\Delta\mathscr{E} = (1 - 2 \times 0.39)(8.314 \times 298/96\,500)\,\ln\,(100/10) = 13\,\text{mV}. \tag{f}$$

The molar fluxes of Na^+, Cl^- and $NaCl$ are all equal and the latter, for example, is obtained as follows. Equation (29) is written for Na^+ and Cl^- and divided by b^+ and b^-, respectively. The results are added and, since the concentrations of Na^+, Cl^- and $NaCl$ are the same, the terms with the potential gradients cancel and we obtain

$$J_1 = - \frac{2b^+ b^-}{b^+ + b^-} \hat{R}T \frac{dc_1}{dx}. \tag{g}$$

Comparing this result with (30) yields the relation between mobilities of ions and the diffusion coefficient for $NaCl$:

$$D_1 = \frac{2b^+ b^-}{b^+ + b^-} \frac{\hat{R}T}{} = 2(1 - t^+)\,b^+\hat{R}T. \tag{h}$$

The mobility of $NaCl$ is obtained directly from (31) as

$$b_1 = 2\,b^+ b^-/(b^+ + b^-) = 2(1 - t^+)b^+. \tag{i}$$

In contrast to $NaCl$, the transference numbers of the ions of KCl are very nearly the same (see, e.g., Harned and Owen, 1958):

$$t^+ = 0.49; \quad t^- = 0.51\,(\text{KCl}), \tag{j}$$

so that b^+, b^- and b_1 are approximately equal and the diffusion potential becomes very small. This fact is used, e.g., in the construction of electrodes for measuring the electrical potential difference between the inside and the outside of cells.

Finally, the generalization of (h) to an electrolyte with arbitrary ionic valencies is

$$D_1 = \frac{b^+ b^- (z^+ - z^-)}{z^+ b^+ - z^- b^-} \hat{R}T. \tag{k}$$

Example 7.4

Each reservoir of the diffusion cell in Figure 7.4 again contains an aqueous solution of NaCl with concentration $c_{1A} = c_{1B} = 100$ mmol/l at 25 °C. An electrical potential difference of 100 mV is now established between the two reservoirs by external means.

Calculate the flux of NaCl and its direction, given $D_1 = 1.48 \times 10^{-9}$ m^2/s.

The problem is the reverse of Example 7.3 in that the externally induced potential difference forces the establishment of a concentration gradient and a transport of matter provided that the mobilities of the ions are different. This follows both from the principle of electroneutrality, according to which the local concentrations of Na$^+$, Cl$^-$ and NaCl must be the same,

$$c^+ = c^- = c_1, \tag{a}$$

and from the movement of charge at constant potential difference, according to which the fluxes of negative and positive ions reaching the anode and cathode, respectively, must be the same:

$$z^+ J^+ = z^- J^-; \quad z^+ = 1, z^- = -1. \tag{b}$$

Because of the convention that fluxes J are positive in the x-direction, the ionic fluxes are oppositely directed. When (29) is written for Na$^+$ and Cl$^-$, respectively, and (a) and (b) are used, elimination of fluxes gives

$$\frac{dc_1}{dx} = -c_1 \frac{b^+ - b^-}{b^+ + b^-} \frac{\mathscr{F}}{\hat{R}T} \frac{d\mathscr{E}}{dx}, \tag{c}$$

which shows the magnitude of the externally induced concentration gradient. It is zero for $b^+ = b^-$, a condition which is nearly true for KCl. The flux of NaCl is obtained by substituting (c) into (30):

$$J_1 = -D_1 c_1 \frac{b^- - b^+}{b^+ + b^-} \frac{\mathscr{F}}{\mathring{R}T} \frac{\mathrm{d}\mathscr{E}}{\mathrm{d}x}. \tag{d}$$

Using (a) from example 7.3, $L = 0.005\,\text{m}$ from Example 7.1, and $D_1 = 1.48 \times 10^{-9}\,\text{m}^2/\text{s}$ for NaCl, (d) gives

$$J_1 = (1.48 \times 10^{-9}) \times 100 \times \frac{1 - 0.39/0.61}{1 + 0.39/0.61} \frac{96\,500}{8.314 \times 298} \frac{0.1}{0.005} \tag{e}$$

$$= 2.54 \times 10^{-5}\,\text{mol/m}^2\text{-s}.$$

Finally, (d) can be generalized for an electrolyte with arbitrary ionic valencies as

$$J_1 = -D_1 c_1 \frac{(z^-)^2\, b^- - (z^+)^2 b^+}{z^+ b^+ - z^- b^-} \frac{\mathscr{F}}{\mathring{R}T} \frac{\mathrm{d}\mathscr{E}}{\mathrm{d}x}. \tag{f}$$

7.2.3 Membrane systems

Biological membranes, which enclose cells and structures within cells, are complex structures that can be described as bimolecular films of phospholipids with inclusions of proteins having special functions, such as chemical pumps (see also Section 7.4). The thickness of membranes is, in general, of the order of 10 nm, i.e. about 2000 times smaller than the diameter of most cells.

The description of transport of matter across cell membranes usually treats the interior of the membrane as a continuous system, while transport between the membrane and intracellular and extracellular solutions is considered to take place between two discrete systems. This description is consistent with the anticipation that the major resistance to transport lies within the membrane, hence that there will be no substantial concentration gradients (polarization) in the solutions on either side.

In the membrane, considered as a continuous system, the usual assumptions are, (i) steady conditions and no flow $J_j = $ constant and $V = 0$ in (33), (ii) $b_j = $ constant, (iii) $T = $ constant, and (iv) a linear gradient of electrical potential, $\mathrm{d}\mathscr{E}/\mathrm{d}x = \Delta\mathscr{E}/L = $ constant, where

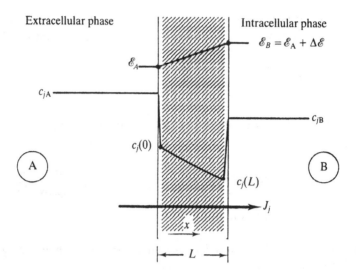

Fig 7.5 Distribution of electrical potential $\mathscr{E}(x)$ and concentration $c_j(x)$ across a membrane without fixed net charges.

$\Delta \mathscr{E} = \mathscr{E}_B - \mathscr{E}_A$ and L is the thickness of the membrane, Figure 7.5. With these assumptions, (29) can be written as

$$dc_j/dx + (A_j/L)c_j = - J_j/D_j, \tag{35}$$

where $D_j = b_j \hat{R} T$ according to (30), and

$$A_j = z_j \mathscr{F} \, \Delta \mathscr{E}/\hat{R}T. \tag{36}$$

Integration of (35) with $c_j = c_j(0)$ at $x = 0$ gives

$$c_j(x) = c_j(0) \, e^{-A_j x/L} - (J_1 L/D_j A_j) \, (1 - e^{-A_j x/L}). \tag{37}$$

Substituting $c_j = c_j(L)$ into (37), the flux becomes

$$J_j = - \frac{D_j A_j}{L} \; \frac{c_j(L) - c_j(0) \exp(-A_j)}{1 - \exp(-A_j)} . \tag{38}$$

The concentrations within the membrane close to its surface are often considerably lower than the concentrations of the adjacent solutions. Usually, this kind of difference in concentration is described by a single *distribution coefficient*, K_m:

$$K_m = c_j(0)/c_{jA} = c_j(L)/c_{jB}, \tag{39}$$

which depends on the properties of the membrane as well as the solute. The value of K_m is usually quite small (typically 0.1–0.001). This is due to the large difference in concentration of hydrogen bonds between the acqueous solutions and the membrane (see, e.g., Weissmann and Claiborne, 1975). Differences in the concentration of ions between the solutions and the membrane may also be due to the presence of fixed net charges in the membrane. In this case, the problem can be treated by assuming local Donnan equilibria at both boundaries with ensuing discontinuities in the electrical potential.

Substituting (39) in (38) gives the flux in terms of the concentrations in the extra- and intracellular phases

$$J_j = -A_j P_j \frac{c_{jB} - c_{jA} \exp(-A_j)}{1 - \exp(-A_j)} \tag{40}$$

where P_j denotes a *permeability coefficient*,

$$P_j = D_j K_m / L, \tag{41}$$

which has the dimension of velocity. Its value depends on the characteristics of membrane and solute and can vary several orders of magnitude, e.g. 10^{-21} m/s for sucrose and 10^{-4} m/s for water, both in the human red cell membrane.

Equation (40) can be generalized to include the effect of a pressure gradient. If we assume, in addition to the previously mentioned assumptions, that $dp/dx = \Delta p_m / L = $ constant, where $p_m = p(0) - p(L)$, (29) again gives (35)–(40) with the only change that

$$A_j = z_j \mathscr{F} \Delta \mathscr{E} / \hat{R}T + \hat{v}_j \Delta p_m / \hat{R}T. \tag{42}$$

Use of this result requires a relation between Δp_m in the membrane and $\Delta p = p_B - p_A$ over the membrane. Such a relation is obtained by assuming local equilibrium for water at each phase boundary (cp. Section 6.4.4) and this introduces *the concept of osmotic pressure in a non-equilibrium membrane system* (Garby, 1957). If water and its dissolved components were considered to be a continuous phase in the pores of the membrane, the transport would then take place not only by diffusion but also by flow caused by an osmotic driving force of the magnitude Δp_m. Note that the assumption $V = 0$ in (33) would not be valid in this case.

The total *electrical current density* (C/s-m^2) through the membrane is the sum of contributions of all fluxes of ion transport:

$$i = \Sigma_{ions} i_j; \quad i_j = J_j z_j, \tag{43}$$

where passive diffusion fluxes J_j of ions as well as active fluxes J_j^p in ionic pumps are included. Note that the transference number for an ionic component j is

$$t_j = i_j/i; \quad \Sigma t_j = 1, \tag{44}$$

which is often used to describe the passive diffusion of charge, cp. Example 7.3.

The result described by (40), valid for $V = 0$, generates a number of special cases.

Firstly, when the membrane potential is zero, or when component j is a non-electrolyte, (36) gives $A_j = 0$, and (40) reduces (e.g. by use of series expansion) to

$$J_j = - P_j (c_{jB} - c_{jA}), \tag{45}$$

which is the phenomenological result (15).

Secondly, in systems with electrolytes, the membrane potential is in general finite. At equilibrium, $J_j = 0$ and (40) reduces to (6.47) for $\Delta p = 0$, from which $\Delta \mathscr{E}$ can be calculated for given differences in concentration (see Example 6.6). At non-equilibrium, $J_j \neq 0$ and $\Delta \mathscr{E}$ can be calculated from (40) and (43) whenever differences in concentration and electrical current are known. If the electrical current is zero, $i = 0$ in (43), and contributions from ionic pumps, J_j^p, are included explicitly, (40) gives

$$\sum_{\text{ions}} -z_j A_j P_j \frac{c_{jB} - c_{jA} \exp(-A_j)}{1 - \exp(-A_j)} + \sum_{\text{ions}} z_j J_j^p = 0, \tag{46}$$

which can be solved iteratively for $\Delta \mathscr{E}$, being a factor in A_j, cp. (36). For monovalent ions, (46) can be rewritten as

$$\Delta \mathscr{E} = \frac{\hat{R}T}{\mathscr{F}} \ln \left\{ \frac{\Sigma^+ P_j c_{Aj} + \Sigma^- P_j c_{Bj} - (\hat{R}T/\mathscr{F}\Delta\mathscr{E}) J_{\text{net}}^p}{\Sigma^- P_j c_{Aj} + \Sigma^+ P_j c_{Bj} - (\hat{R}T/\mathscr{F}\Delta\mathscr{E}) J_{\text{net}}^p} \right\}, \tag{47}$$

where Σ^+ and Σ^- denote summation over cations and anions, respectively, B refers to intracellular and A to extracellular fluid (Figure 7.5), and $J_{\text{net}}^p = \Sigma z_j J_j^p$ denote the net pump flux of ions from A to B. In the case of passive diffusion only, and when the pump is electrically neutral ($J_{\text{net}}^p = 0$), equation (47) can be solved explicitly and reduces to the *Hodgkin–Katz–Goldmann equation*:

$$\Delta \mathscr{E} = \mathscr{E}_B - \mathscr{E}_A = \frac{\hat{R}T}{\mathscr{F}} \ln \left\{ \frac{\Sigma^+ P_j c_{jA} + \Sigma^- P_j c_{jB}}{\Sigma^- P_j c_{jA} + \Sigma^+ P_j c_{jB}} \right\}. \tag{48}$$

7.3   Irreversible thermodynamics

The preceding section showed that explicit expressions for the production of entropy can be obtained by the use of mass conservation, the first law and the Gibbs relation, cp. e.g. (3), derived from (1) and (2). The resulting expression for the entropy production is of the form

$$entropy\ production\ =\ driving\ force \times flux \qquad (49)$$

or, with several irreversible phenomena, a sum of contributions:

$$\dot{\sigma} = \Sigma X_j\,J_j \geqslant 0, \qquad (50)$$

where the driving force X_i is the gradient of, or the difference in, potential, an intensive property, and the flux J_i is the associated transferred extensive property.

The theory of irreversible thermodynamic processes, see, e.g., de Groot and Mazur (1962) and Glansdorff and Prigogine (1971), contains *inter alia* a systematic treatment of expressions for resulting fluxes, i.e. constitutive relations for transport processes. At equilibrium, the entropy production is zero, $\dot{\sigma} = 0$, because all processes cease, $J_i = 0$, and the state of the system is uniform and at thermodynamic equilibrium, $X_i = 0$. For small deviations from equilibrium, the flux can be expected to depend linearly on the associated driving force. This dependency must be homogeneous because the flux must vanish when the driving force becomes zero at equilibrium.

These considerations form the basis for the postulate of the theory of *linear phenomenological relations*:

$$flux\ =\ phenomenological\ coefficient \times driving\ force \qquad (51)$$

or, explicitly,

$$J_i = \sum_k L_{ik} X_k = L_{i1} X_1 + \ldots + L_{ii} X_i + \ldots, \qquad (52)$$

which implies that the flux of component i is influenced by all driving forces X_k. Terms with L_{ik}, where $i \neq k$, are cross-effects, while terms with L_{ii} are primary effects.

Expression (50) identifies conjugate pairs of forces and fluxes, X_i and J_i. Substitution of (52) into (50) gives

$$\dot{\sigma} = \Sigma_i\,\Sigma_k\,L_{ik}\,X_k\,X_i \geqslant 0, \qquad (53)$$

from which it follows that phenomenological coefficients, ultimately to be determined by experiment, are subject to restrictions

$$L_{ii} \geqslant 0; \quad 4 \, L_{ii} \, L_{kk} - (L_{ik} + L_{ki})^2 \geqslant 0. \tag{54}$$

Using fluxes and forces described by (50), the *Onsager reciprocity relation* takes the form:

$$L_{ik} = L_{ki}, \tag{55}$$

which is based on theoretical considerations of microreversibility at the molecular scale.

The summation in (52) includes all phenomena k which have the same tensor rank (or differ in rank by two) as the primary phenomenon i. This follows from *Curie's symmetry principle*, which is derived from the rules of transformation of tensor relations, and which guarantees coordinate invariance of the resulting constitutive relations. In this text, we deal only with phenomena with tensor rank zero (scalar phenomena) and one (vectorial phenomena). Phenomena of tensor rank two include the stress tensor associated with viscous flow, electrostriction and magnetostriction. For each class of coupled irreversible phenomena, (53) must be fulfilled, irrespective of possible irreversible phenomena of other classes.

Scalar phenomena, here as chemical reactions, take place at local non-equilibria (Section 7.1). It follows from (52) that fluxes (reaction rates) ξ_i must be expected to depend on all forces (affinities) $-A_k/T$, cp. (5). However, in most cases cross-effects prove to be small compared to primary effects. In any case, the linear form of (52) requires small deviations from equilibrium.

Vector phenomena occur at spatial non-equilibria, both in discrete systems (Section 7.2.1) and in continuous systems (Sections 7.2.2 and 7.2.3). In the last-mentioned case, it follows from (52) that one-dimensional fluxes in the x-direction (heat conduction, mass diffusion of each component j in a mixture, and transport of electrical charge) must be expected to depend on all existing forces (gradients in temperature and electrochemical potential, i.e. pressure, concentration, and electrical potentials). Thus in a one-component system, a temperature gradient will only cause heat transfer, while, in a binary system, it will also cause mass transfer. This cross-effect (thermo-diffusion or *Soret effect*) can be used to separate components in a binary gas mixture. In a similar fashion, a concentration gradient contributes to the transport of heat (diffusion-thermo or *Dufour effect*). However, in biological systems such cross-effects are usually negligible compared to the primary effects.

It must be stressed that the thermodynamic couplings between irreversible phenomena in (52) are constitutive, since they appear in the constitutive relations. However, another type of coupling, by way of general laws, say mass and energy conservation, can of course also take place. In a process with combined chemical reaction and diffusion of matter, for example, both rates depend on the chemical potential, cp. (4) and (25), but there is no constitutive coupling here since one phenomenon is scalar and the other vectorial. The coupling occurs through the conservation of mass which involves both the mass transfer of reactants and products to and from the locality of chemical reaction and the rate at which this reaction consumes or produces these components.

Finally, we note that it is sometimes useful to introduce *generalized forces*, defined by

$$F_i = T X_i. \tag{56}$$

When (50) is multiplied by the absolute temperature, we obtain the *irreversibility power* cp. (5.9)

$$\dot{I} = T \dot{\sigma} = \Sigma F_i J_i \geqslant 0, \tag{57}$$

which equals the energy per unit of time dissipated through irreversibilities. Optimal operation of a system, including the criteria and mechanisms for its control, ought to be such that $\dot{I}$ is minimized (Glansdorff and Prigogine, 1971). For the diffusion of matter in an isothermal system, (21) and (25) imply that $F_j = -(d\bar{\mu}_j/dx)$ and $J_j = -b_j c_j (d\bar{\mu}_j/dx)$.

7.4   Model of membrane transport in cells

Maintenance of the very complicated, often enzymatically catalyzed, chemical processes in most cells requires, among other things, (i) concentrations of proteins (in particular enzymes) and organic phosphate, both negatively charged at the actual pH level (about 7.2), which are much higher than in the solutions surrounding the cells; and (ii) net transport through the cell membrane of a number of small molecules (glucose, amino acids, etc.) needed for the chemical processes.

These requirements are fulfilled by the cell membrane being impermeable to the anions but permeable to the small molecules mentioned. As a consequence, the membrane is also permeable to water and to small ions (e.g. Na^+, K^+, Cl^- and HCO_3^-). However, the presence of impermeable ions, even at very small concentrations, will give rise to a significant osmotic pressure and, since the cell membrane cannot withstand pressure differences of more than a few millimetres of water, the system must arrange for active transport of at least one of the permeating components from the cell to the surroundings. This transport requires an input of energy (free energy) and must be regulated in such a manner that the chemical potential of water (or the total activity of the solutes) becomes the same on both sides of the membrane.

The requirement of such transport is generally fulfilled by the presence in the cell membrane of chemical pumps for the active transport of sodium ions out of the cells. These pumps, which are driven by free energy from the reaction $ATP \rightarrow ADP + P$, usually operate with a stoichiometric coupling and simultaneous transport of potassium ions in the opposite direction, into the cells. The coupling can take place with a coupling ratio $1:1$, in which case the pump is electrically neutral, or, more often, with a ratio $3:2$, in which case the pump is electrogenic.

An important consequence of the active transport of ions is the generation and maintenance of an electrical potential difference across the membrane, with the inside of the cell negative, the so-called *resting potential*. This situation is not only a consequence of necessary control of the pressure, but it is also a requirement for the generation of controlled variations in this potential difference, so-called *action potentials*, and for the associated transmission of electrical signals in nerves and in muscle cells.

Given (i) the composition of the surrounding medium, (ii) the concentration and electrical charge of impermeable ions in the cell, (iii) electroneutrality in cell and surrounding medium and (iv) the coupling ratio of the pump, the rate of active transport of sodium must be adjusted so that, at isothermal conditions, there is thermodynamic equilibrium for all permeable components that are not transported actively.

These considerations can best be illustrated by performing calculations on a specific example for the state and transport processes in a cell. Due to the complexity of the real situation and lack of data on several details, the following example is simplified considerably.

Example 7.5

The cell and its surroundings in Figure 7.6 are assumed to be characterized by the following concentrations:

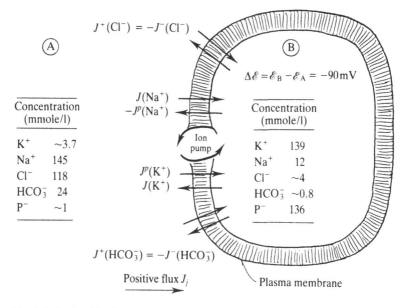

Fig 7.6 A simplified model of membrane transport of ions for a cell with typical values for concentrations and electrical potential.

Component	K^+	Na^+	Cl^-	HCO_3^-	P^-	
(A) extracellular mmol/l	3.7	145	118	24	~1	(a)
(B) intracellular mmol/l	139	12	~4	~0.8	136	(b)

where P^- denotes anions of protein and organic phosphate, to which the membrane is impermeable. The mean value of the charge on P^- is -6.7 and -1.08 in (A) and (B), respectively, so that electroneutrality is fulfilled for both phases. An electrical potential difference of $\Delta\mathscr{E} = \mathscr{E}_B - \mathscr{E}_A = -90\,\mathrm{mV}$ has been measured. Pressure and temperature is 1 atm. and 310 K, respectively, and activity coefficients for any component inside and outside the cell are assumed to be the same.

In the following, diffusion fluxes are counted positive in the direction from the surroundings A to the interior of the cell B, J^+ denotes the unidirectional flux in this direction and J^- the corresponding flux in the opposite direction.

Using tracers and observation of initial fluxes (cp. Example 7.2), the uni-directional fluxes have been determined to

Component	K^+	Na^+	Cl^-	HCO_3^-	P^-	
$10^{10} \times J^+$ (mol/m^2-s)	–	250	14.2	3.6	0	(c)
$10^{10} \times J^-$ (mol/m^2-s)	722	–	–	–	0	(d)

Such results allow us to calculate the effective permeability coefficient P_j in (40). For $J^+(Na^+)$, we obtain from (36)

$$A(Na^+) = (+1) \times 96\,500 \times (-0.090)/(8.314 \times 310) = -3.37, \quad (e)$$

and, from (40), since the concentration of tracer at the receiving side of the membrane is zero (cp. Example 7.2),

$$P(Na^+) = \frac{J^+(Na^+)\,\{1 - \exp[-A(Na^+)]\}}{A(Na^+)\,c_A(Na^+)\,\exp[-A(Na^+)]}, \quad (f)$$

which, on substitution of values, gives

$$P(Na^+) = \frac{2.5 \times 10^{-8} \times [1 - \exp(3.37)]}{(-3.37) \times 145 \times \exp(3.37)} = 4.94 \times 10^{-11}\,\text{m/s}. \quad (g)$$

Correspondingly, we find for K^+ that $A(K^+) = A(Na^+) = -3.37$, and since $C_A(K^+) = 0$ for tracer

$$P(K^+) = \frac{7.22 \times 10^{-8}\,[1 - \exp(3.37)]}{(-3.37) \times 139} = 4.33 \times 10^{-9}\,\text{m/s}. \quad (h)$$

Hereafter, net fluxes for diffusion of Na^+ and K^+ in and out of the cell, respectively, can be calculated from the complete form of (40) as

$$J(Na^+) = 3.37 \times (4.94 \times 10^{-11}) \times \frac{12 - 145 \times \exp(3.37)}{1 - \exp(3.37)}$$

$$= 2.49 \times 10^{-8}\,\text{mol/m}^2\text{-s}, \quad (i)$$

$$J(K^+) = 3.37 \times (4.33 \times 10^{-9}) \times \frac{139 - 3.7 \times \exp(3.37)}{1 - \exp(3.37)}$$

$$= -1.63 \times 10^{-8}\,\text{mol/m}^2\text{-s}. \quad (j)$$

We note that result (j) is critically dependent on the value of the concentration of K^+ in the outside medium. The minus sign in (j) signifies that the net flux is positive from B to A, i.e. out of the cell. Assuming steady state for all processes we conclude that the chemical pump must deliver an active flux of Na^+ out of the cell and, simultaneously, an active flux of K^+ into the cell, amounting to

$$J^P(Na^+) = -2.49 \times 10^{-8}\,\text{mol/m}^2\text{-s}, \tag{k}$$

$$J^P(K^+) = +1.63 \times 10^{-8}\,\text{mol/m}^2\text{-s}, \tag{l}$$

so that the coupling ratio for $Na^+ : K^+$ is $1.53 = 3:2$, and the pump is electrogenic.

By performing calculations (e)–(j) for anions and using the unidirectional fluxes in (c), the net fluxes of chloride and bicarbonate ions are found to be negligible. The conclusion is that the transport of anions across the membrane is passive, as presumed for the stationary condition in Figure 7.6. Values for permeability coefficients and diffusion fluxes of all components are shown below.

Component	K^+	Na^+	Cl^-	HCO_3^-	P^-	
$10^{10} \times P_j$ (m/s)	43.3	0.494	1.0	1.25	0	(m)
$10^{10} \times J_j$ (mol/m²-s)	163	249	0.2	0.1	0	(n)

From the values in (n) and (c) for chloride and bicarbonate ions and considering the uncertainty in the values (a)–(c), we conclude that these ions are in diffusion equilibrium.

The osmotic pressure difference across the membrane is calculated according to (6.50) (cp. Example 6.5) using concentrations in (a) and (b):

$$P_B - P_A = 0.0827 \times 310 \times [(139 + 12 + 4 + 0.8 + 136)$$
$$- (3.7 + 145 + 118 + 24 + 1)] \times 10^{-3} = 0.0025\,\text{atm.} \tag{o}$$

This corresponds to about 25 mm water, a value which must be considered in relation to the uncertainty of the measured concentrations (the result corresponds to an osmotic activity difference of only 0.1 mmol/l).

In order to illustrate the consistency of the model, we now calculate the electrical potential difference across the membrane and compare it with the observed value of $-90\,\mathrm{mV}$. We first note that the steady conditions require that the sum of all passive diffusion fluxes and pump fluxes for each component be zero, cp. (k), (l) and (n). Each contribution i_j in (43) is therefore zero and the total charge current across the membrane is zero. The membrane potential can therefore be calculated in several ways.

For each anion there is diffusion equilibrium, and $J_j = 0$ in (40) reduces to (6.47) with the first term equal to zero, i.e. (b) in Example 6.6 for chloride ions. Using $\hat{R}T/\mathscr{F} = 8.314 \times 310/96\,500 = 0.0267\,\mathrm{V} = 26.7\,\mathrm{mV}$, this gives

$$\Delta\mathscr{E} = \mathscr{E}_B - \mathscr{E}_A = 26.7 \times \ln(4/118) = -90\,\mathrm{mV}. \tag{p}$$

Correspondingly, for bicarbonate ions,

$$\Delta\mathscr{E} = 26.7 \times \ln(0.8/24) = -90.8\,\mathrm{mV}. \tag{q}$$

These results confirm the equilibrium states for these ions.

In the case of cations, there is diffusion non-equilibrium, but the ion pump compensates so that the total charge current becomes zero. Therefore, (46) can be used and written in the form

$$J(\mathrm{Na}^+) + J(\mathrm{K}^+) + J^p(\mathrm{Na}^+) + J^p(\mathrm{K}^+) = 0. \tag{r}$$

When the coupling ratio, n^p, is known:

$$n^p = J^p(\mathrm{Na}^+)/J^p(\mathrm{K}^+) = -1.53, \tag{s}$$

the last two terms in (r) can be combined to $(1 + n^p)J^p(\mathrm{K}^+)$ and, using $J^p(\mathrm{K}^+) = -J(\mathrm{K}^+)$, (r) reduces to

$$J(\mathrm{Na}^+) - n^p\, J(\mathrm{K}^+) = 0. \tag{t}$$

Using (40) and (36), we obtain

$$\Delta\mathscr{E} = \frac{\hat{R}T}{\mathscr{F}}\, \ln\left\{ \frac{P(\mathrm{Na}^+)\, c_A(\mathrm{Na}^+) + n^p P(\mathrm{K}^+)\, c_A(\mathrm{K}^+)}{P(\mathrm{Na}^+)\, c_B(\mathrm{Na}^+) + n^p P(\mathrm{K}^+)\, c_B(\mathrm{K}^+)} \right\}, \tag{u}$$

which, using values from (a), (b), (m) and (s), gives

$$\Delta\mathscr{E} = -90.02\,\mathrm{mV}. \tag{v}$$

This result only shows that the values in the model are consistent. The calculation presumes that the coupling ratio of the pump is known (calculated here on the basis of an observed potential difference of $-90\,\mathrm{mV}$). In general, the resting potential of a cell cannot be calculated on the basis of observed concentrations and permeabilities alone. It should also be noted that systematic methodological errors (in particular with respect to the concentration of K^+ in A here) and assumptions of ideal solutions, as well as linear electrical potential gradient in (40), can cause considerable difficulties in characterizing states and processes in cells.

As a final note, we repeat and stress that the cell described in Example 7.5 represents a considerable simplification of the real situation, e.g. by the choice of concentrations which give vanishing net fluxes of anions. Real cells also have, in addition to the transports described here, chemical pumps for transport of calcium and protons out of the cells. Furthermore, several phenomena with respect to formation (from carbon dioxide) and transport of bicarbonate, having a much higher concentration in the cells than given here, have not been considered in the example.

8

Expenditure of exergy (exergy balance)

The first law is a statement concerning quantities of energy. It requires that the sum of all forms of energy is conserved during any process, even though energy is transformed from one form to another. The second law is concerned with changes in entropy. It is qualitative in stating whether a given process will proceed spontaneously. It is quantitative in stating how far the process will deviate from the ideal reversible process. This deviation is determined by the entropy production of the process or its irreversibility, which is calculated by using the first and the second laws, as well as the Gibbs relation.

These considerations can be generalized and quantified by introducing the concept of useful energy (exergy or availability). The concept defines an extensive state property (in essence the Gibbs energy) that describes the quality of the energy. Quantitatively, the exergy belonging to a given energy state of a material system, and given surroundings, is the maximal amount of energy that is at the system's disposal for use. The decrease in exergy of a system that undergoes a process from one state to another is equal to the expenditure of useful energy during the process. This change can be fully used in an ideal (reversible) process, but it can only be partially used in a real (irreversible) process since some of the exergy will be destroyed in irreversibilities.

These ideas are expressed in the exergy balance which describes how the quality of the energies in a given process changes, e.g. due to transformation and expenditure of useful energy. The exergy balance will be derived in both a general form and in a simplified form that is useful for the description of processes in living systems. The concepts of exergy transformation and exergy efficiency are introduced, and their use is illustrated by examples. Finally, a simplified model of a living system is analysed.

222

8.1 Useful energy

The concept of useful energy is conveniently explained by reference to a work producing device (the power station) that delivers mechanical power $\dot{W}$ by expenditure of heat power $\dot{Q}_H$ from a reservoir at high temperature T_H (the boiler). Experience shows that this is possible only when a certain heat power $\dot{Q}_L$ is removed and transferred to another reservoir at a lower temperature T_L (the surroundings). The study of processes of this kind motivated the beginnings of theoretical considerations that led to classical thermodynamics.

The necessity to remove a certain heat energy δQ_L to the surroundings when a cyclic process is delivering mechanical work δW with transformation of added heat δQ_H can be illustrated and rendered obvious as follows (Figure 8.1).

The air below the piston in a cylinder (e.g. a bicycle pump) is heated with the piston in a fixed position (process $1 \to 2$, added δQ_H). Thereby, both temperature and pressure increase so that work can subsequently be delivered to the surroundings while the piston rises and the air expands to the lower, initial pressure $p_3 = p_1$ (adiabatic process $2 \to 3$, removed work δW). In order to return to the initial state, the air must be cooled so that the volume decreases and the piston falls at constant pressure (process $3 \to 1$, removed heat δQ_L).

If this cycle is repeated many times, the time-averaged result over a long time period can be interpreted in terms of the steady process

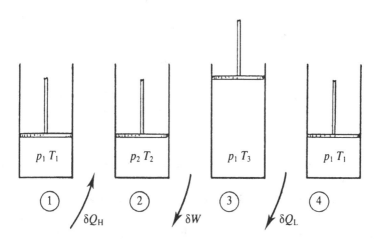

Fig 8.1 Cyclic process consisting of three separate processes for the air enclosed in a cylinder with piston.

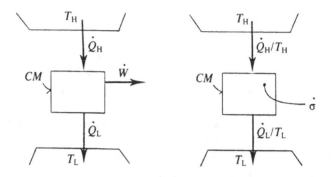

Fig 8.2 The first and second law for the control mass consisting of a machine that transforms heat power into work power in a steady (or equivalent cyclic) process.

considered in Figure 8.2. The individual processes can be readily analysed and the work calculated, say if the air is assumed to be an ideal gas (Section 3.8.5) and the initial state (T_1, p_1) and p_2, or T_2 are specified, cp. Example 4.6.

The first and second laws, ((4.1)) and ((5.1)), for steady processes in the control mass in Figure 8.2 give (with positive powers in accordance with the arrows)

$$0 = \dot{Q}_H - \dot{Q}_L - \dot{W}; \quad 0 = \dot{Q}_H/T_H - \dot{Q}_L/T_L + \dot{\sigma}. \tag{1}$$

Since $\dot{\sigma} \geq 0$, it follows that $\dot{Q}_L \geq (T_L/T_H)\, \dot{Q}_H$, so that at least the fraction T_L/T_H of the added heat power $\dot{Q}_H$ must be removed as heat $\dot{Q}_L$ and only the remaining part of $\dot{Q}_H$ can be transformed into work:

$$\dot{W} = \dot{Q}_H - \dot{Q}_L \leq (1 - T_L/T_H)\, \dot{Q}_H. \tag{2}$$

From this it follows that the efficiency $\eta = \dot{W}/\dot{Q}_H$, cp. (4.14), can be maximally equal to the *Carnot efficiency* for a work producing machine:

$$\eta_C = 1 - T_L/T_H. \tag{3}$$

It should be noted that for a heat pumping machine (heat pump, refrigerator), all arrows in Figure 8.2 would be reversed and all energy flows would change sign. The ideal machine would then be one that used the least work power for a given delivered heat power (heat pump) or removed heat power (refrigerator). Recall the sign convention of work, being positive when supplied to a system. The minimum supply of a positive quantity (work supplied) is the maximum yield of a negative quantity (work extracted). This is readily shown on a coordinate axis for W by considering positive and negative values.

The above example shows that heat energy, in addition to its *quantity* i.e. the amount of energy Q_H, can also be described by its *quality*, i.e. its usefulness (the exergy). The measure for the usefulness is the amount of energy W in the form of work that can be extracted from Q_H, given the details of the process, here T_H and T_L where T_L represents the surroundings.

8.1.1 Definition

The useful energy (exergy, availability, see, e.g., van Wylen and Sonntag, 1965; Moran, 1982) associated with a given form of energy, is *defined* as the maximal energy in the form of work that can be extracted from the energy, given some surroundings that characterize a state of rest, and is a reservoir for heat exchange at temperature T_0. In this context, work is to be understood broadly as the highest energy quality: mechanical work, electrical work and chemical work (Gibbs free energy).

The term 'surroundings' is to be understood as a reference state of rest, relevant for the given problem. For a power station, it is the surrounding nature with its cooling water; for a living system as a whole, the surroundings; for part of a living system, the remaining, often isothermal, parts. The exergy is therefore *a state property that depends on the chosen surroundings*.

We first determine the exergy for various forms of energy associated with a given state. Then (in Section 8.2) we derive the exergy balance which is needed to determine the change in exergy for a process with given initial and final states.

8.1.2 Energy in transit

From the definition of exergy it follows that the exergy ψ_W associated with work energy W is equal to the work itself:

$$\psi_W = W. \tag{4}$$

This is also true for the previously introduced chemical work W_{Ch}, which appears as transferred internal chemical energy between coupled reactions:

$$\psi_{W,Ch} = W_{Ch}. \tag{5}$$

The extent to which the useful energy released by one reaction is also utilized in the other reaction to which it is coupled depends on the details of the coupling.

Next, the example in Figure 8.1 shows that the exergy ψ_Q associated with heat energy for disposal at absolute temperature T is

$$\psi_Q = (1 - T_0/T)Q. \tag{6}$$

This follows from (2) when the temperature of the surroundings is T_0 ($< T$). It also follows that $\psi = 0$ for $T_0 = T$, but under such isothermal conditions, $Q = 0$, of course. The heat Q in (6) includes radiation energy associated with thermal radiation. On the other hand, the radiation energy Q_u is useful energy because it is utilized in photosynthesis with direct conversion into internal chemical energy (see Section 5.4.2), and we have

$$\psi_{Q_u} = Q_u. \tag{7}$$

While heat energy is central to most man-made energy conversion systems, such as power stations, refrigerators, etc., it is of no relevance in bioenergetics because living systems are unable to convert thermal energy into useful forms of work (see Examples 8.2 and 8.5).

8.1.3   Pure substance. Thermomechanical exergy

Energy forms associated with matter in the forms of mass or mass flow are summarized in Figure 3.11. The corresponding exergy is calculated as the work energy that is removed when the matter is brought to equilibrium with the surroundings at a state of rest in a reversible process in which there is an exchange of heat energy solely with the surroundings at T_0.

Use of the first and the second laws, or the exergy balance (Section 8.2) gives, for a control mass M,

$$\psi_M = M[(u - T_0 s + p_0 v + V^2/2 + \varphi) - (u_0 - T_0 s_0 + p_0 v_0 + \varphi_0)] \tag{8}$$

and for a mass flow $\dot{m}$,

$$\psi_{\dot{m}} = \dot{m} [(h - T_0 s + V^2/2 + \varphi) - (h_0 - T_0 s_0 + \varphi_0)], \tag{9}$$

where the subscript $(\ )_0$ refers to values of state properties in a state of rest, where the kinetic energy is zero. Results (8) and (9) imply that the imagined reversible process proceeds until the matter has achieved thermal and mechanical equilibrium with the surroundings.

8.1.4   Mixtures. Thermal–mechanical–chemical exergy

In the case of chemical compounds, mixtures and reactive mixtures, the exergy is calculated as for pure matter. It is the work energy that is removed when the matter is brought to thermal, mechanical and electrochemical equilibrium with the surroundings at a state of rest by a reversible process in which there is exchange of heat energy solely with the surroundings at T_0.

Use of the exergy balance (Section 8.2) for a mixture of $N = \Sigma \, N_j$ moles in a control mass gives

$$\psi_N = \Sigma N_j \, (\hat{u}_j - T_0 \hat{s}_j + p_0 \hat{v}_j + (\hat{K}E)_j + \hat{\varphi}_j)$$
$$- \Sigma N_{j0} \, (\hat{u}_{j0} - T_0 \hat{s}_{j0} + p_0 \hat{v}_{j0} + \hat{\varphi}_{j0}), \tag{10}$$

and for a molar flow $\dot{n} = \Sigma \, \dot{n}_j$ of a mixture:

$$\psi_{\dot{n}} = \Sigma \dot{n}_j \, (\hat{h}_j - T_0 \hat{s}_j + (\hat{K}E)_j + \hat{\varphi}_j) - \Sigma \dot{n}_{j0} \, (h_{j0} - T_0 \hat{s}_{j0} + \hat{\varphi}_{j0}). \tag{11}$$

In this case, the composition (N_j or $\dot{n}_j$) of the given mixture is known, while the resulting composition (N_{j0} or $\dot{n}_{j0}$) at a state of rest must be determined from the conditions at thermo-mechanical-chemical equilibrium (Chapter 6).

Such calculations can be complicated, but it helps to consider the equilibrium state as being reached in two steps. During combustion, for example, there is, first, an oxidation of the fuels to chemical equilibrium at (T_0,p_0) so that the law of mass action (6.59) is fulfilled. Since the surroundings contain oxygen, complete oxidation can always take place if the condition for equilibrium requires this. Next, the condition for electro-chemical equilibrium (6.25) must be fulfilled for the individual components of the mixture. This requires, e.g. for gases, that they expand to the respective partial pressures that exist in the surroundings, see, e.g., Moran (1982).

For living systems, the most important contribution comes from the combustion of fuels in chemical reactions whose equilibrium state can be assumed to be fully displaced towards the products. In addition, changes in kinetic and potential energy can be neglected and the pressure is normally the same as that of the surroundings. For subsystems in the internal parts of a living system, the condition $T_0 = T$ can often be used. Otherwise, temperature differences should be taken into account.

Often, we need to calculate changes in exergy from one known state (e.g. food) to another known state (e.g. excrement). For a steady process, the contributions from the state of rest cancel since they are the same for both states. In addition, if changes in kinetic and potential energy are neglected, the change in exergy for a steady molar flow can be written as

$$\psi_{\dot{n}1} - \psi_{\dot{n}2} = [\Sigma \dot{n}_j \, (\hat{h}_j - T_0 \hat{s}_j)]_1 - [\Sigma \dot{n}_j \, (\hat{h}_j - T_0 \hat{s}_j)]_2. \qquad (12)$$

If the process is also isothermal at temperature $T = T_0$, (12) implies that the change in exergy only involves changes in chemical potentials for the components of the mixture.

8.2   Balance of exergy

This general law is a useful supplement to the first and second laws, from which it is derived. We first derive its general form, and then give a simpler form which is relevant and sufficient for biological systems. The difference between the two forms concerns the choice of reference temperature.

The exergy balance in its traditional form is based on the heat exchange solely at the temperature of the surroundings T_0. It is relevant for power stations and industrial processes, etc., which utilize thermal energy directly or indirectly in processes of energy transformation for the production of mechanical or electrical power. The efficiency of these processes depends essentially on the temperature of the surroundings, since the processes demand the removal of heat to the surroundings, cp. Figure 8.2.

In living systems, the situation is quite different. Warm-blooded (homeothermic) animals have established – through evolution – a body temperature that is slightly above the mean temperature of the earth. This appears to be of advantage for complex biochemical processes and for the stability of many biological molecules and aggregates of mole-cules, and it facilitates the necessary heat removal. Neither warm-blooded nor cold-blooded (poikilothermic) animals use processes of energy conversion that are based on two heat reservoirs. All the supply of energy and exergy is associated here with the chemical energy in matter, and with the radiation used in photosynthesis.

It is therefore relevant to base an exergy balance on the temperature of the organism itself. This choice implies that thermal energy is not directly involved in this exergy balance.

8.2.1   General form

For an arbitrary process in a control volume, the exergy balance is derived by subtracting the entropy balance ((5.6)), multiplied by the

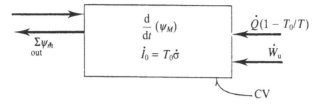

Fig 8.3 The exergy balance for a control volume.

absolute temperature of the surroundings T_0, from the first law ((4.7)), which yields

$$\mathrm{d}[M(u - T_0 s + V^2/2 + \varphi)]/\mathrm{d}t + \Sigma_{\text{out}}\, \dot{m}(h - T_0 s + V^2/2 + \varphi)$$
$$= \dot{Q}(1 - T_0/T) + \dot{W} - \dot{I}_0, \tag{13}$$

where $\dot{I}_0 = T_0\,\dot{\sigma}$ represents the irreversibility power referred to T_0.

If $(h_0 - T_0 s + \varphi)$ multiplied by the conservation of mass ((3.9)) is subtracted from (13), we can rewrite the latter and obtain the *exergy balance* (Figure 8.3):

$$\mathrm{d}(\psi_M)/\mathrm{d}t + \Sigma_{\text{out}}\, \dot{m}\, \psi_m = \dot{Q}(1 - T_0/T) + \dot{W}_{\text{u}} - \dot{I}_0, \tag{14}$$

where expressions (8) and (9) have been introduced, and where

$$\dot{W}_{\text{u}} = \dot{W} - (-p_0\, \mathrm{d}\mathcal{Y}/\mathrm{d}t) \tag{15}$$

denotes the useful work power which must exclude work associated with the change in the control volume against the pressure of the surroundings. The latter contribution is normally of no interest for biological systems since the volume $\mathcal{Y}$ is constant.

According to (14), the temporal increase of exergy within the control volume plus the net outflow of exergy is equal to the exergy supplied by heat and work powers minus the destruction of exergy, determined by the irreversibility of the process.

In rewriting (13) and (14), we use the identity $h_0 = u_0 + p_0 v_0$ and (15) with $\mathcal{Y} = Mv$. Considering the reversible and steady process in Figure 8.4, it can be seen that (13) becomes (9) since $\psi_m = -\dot{W}$. The same result is obtained by (14), in that $\psi_{m,0} = 0$ at a state of rest. In order to derive (8) from (13), we consider a reversible process for a control mass and use (15).

8.2.2 Biological systems

For mixtures and chemical reactions it is convenient to use molar- instead of mass-based specific state properties. In biological systems,

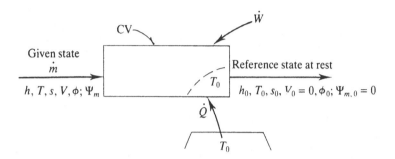

Fig 8.4 Reversible steady process which brings the mass flow from a given state to a state of rest.

changes in kinetic and mechanical potential energy, as well as thermal energy exchange, are usually of no importance so that (13) reduces to

$$d[N(\bar{u} - T\bar{s})]/dt + \Sigma_{\text{out}} \dot{n}(\bar{h} - T\bar{s}) = \dot{Q}_{\text{u}} + \dot{W} + \dot{W}_{\text{Ch}} - \dot{I}, \qquad (16)$$

where $\bar{u}$ and $\bar{h}$ denote electrochemical energy and enthalpy, respectively, cp. (3.133) and (3.134), $N\bar{u}$, $\dot{n}\bar{h}$ and $\dot{n}\bar{s}$ represent summations of contributions in the case of mixtures. Also, exergy contributions such as (5) and (7) are shown explicitly such that $\dot{W}$ does not include chemical work, and $\dot{I} = T\dot{\sigma}$, the reference temperature being that of the system itself.

Expression (16) is seen to involve the Helmholtz function $\bar{a} = \bar{u} - T\bar{s}$ in the first term and the Gibbs function $\bar{g} = \bar{\mu} = \bar{h} - T\bar{s}$ in the second term, both including contributions from electrical potential energy. For most biological applications, we have that $\hat{u} \approx \hat{h}$, cp. (4.44), so that (16) reduces to

$$d(N\bar{\mu})/dt + \Sigma_{\text{out}} \dot{n}\,\bar{\mu} = \dot{Q}_{\text{u}} + \dot{W} + \dot{W}_{\text{Ch}} - \dot{I}. \qquad (17)$$

Result (17) shows that the electrochemical potential $\bar{\mu}$, the Gibbs free energy, effectively represents the useful energy, i.e. the exergy, under the stated conditions.

Finally, (17) can be simplified for isothermal and isobaric reactions without changes in electrical potential energy. As for the derivation of (4.61) and (5.27), we can use the mole balance ((3.73)) with the source term ((3.80)) to rewrite (17) as

$$\Sigma_r \xi_i \,\Delta\hat{g}_{ri} = \dot{Q}_{\text{u}} + \dot{W} + \dot{W}_{\text{Ch}} - \dot{I}, \qquad (18)$$

where the reaction Gibbs function for reaction i, $i = 1,2,\ldots,r$, is given by (5.30), or

$$\Delta \hat{g}_{ri} = \Sigma\, \nu_{ij}\, \hat{\mu}_j. \tag{19}$$

As always, $\dot{I} \geqslant 0$, where the equality sign refers to the ideal, reversible process.

The importance of (18), when all quantities are positive, is as follows. The net outflow of exergy (free energy) from the control volume is equal to the supplied exergy in the form of useful radiation and work minus the destruction of exergy within the control volume due to irreversibility of processes. Similar descriptions can readily be given when the quantities in (18) are not all positive. A negative $\dot{W}$, for example, means that useful work power is removed from the system.

As for (4.61) and (5.27), (18) is valid for cases with isobaric and isothermal reactions, even in cases where there is accumulation (deaccumulation) of components in the control volume. However, assumption of stationarity of intermediary products that remain in the control volume, cp. Figure 4.12, is necessary. The summation on the left-hand side of (18) must therefore include contributions from all reactions which are either non-steady or for which one or more components enter or leave the control volume.

8.3 Expenditure of exergy

Considerations of form, conversion and efficiency of energy in Section 4.3 can be applied to the exergy. This is true even though the exergy, in contrast to the energy, is not conserved but is partly destroyed as determined by the irreversibility of the processes. Such considerations include identification of forms of exergy conversion, total exergy turnover (expenditure), and efficiencies of exergy.

The exergy balances ((14)) or ((16)), applied to a given process in a control volume, can thus be used to identify the forms of exergy that are transformed and to calculate the losses associated with the process.

In a thermal power station, for example, the exergy associated with internal chemical energy in a fossil fuel is converted into heat energy at high temperature and this is then transformed into mechanical work, which is exergy per definition. The destruction of exergy in these processes is considerable so that the useful work energy is normally less than half of the initial chemical exergy.

In living systems, chemical energy in foodstuffs, supplied from the surroundings or taken from depots, is transformed into chemical exergy in ATP synthesis. This is then either transformed into chemical exergy in other synthetic processes, or used to fuel 'pumps' for maintenance of concentration gradients or to create tension in muscles for the performance of internal or external mechanical work. Each step of transformation invariably takes place with a loss of exergy.

8.3.1 Energy expenditure in biological systems

Analogous to the concept of energy expenditure (Section 4.3.2), it is useful to introduce the concept of exergy expenditure as a measure of the thermodynamic costs for maintenance of the processes in the whole – or in parts – of a living system. This expenditure of exergy can be deduced from the exergy balance when the functional conditions of the system have been specified.

We specify *normal conditions* as those where (i) the system is in a steady state or expends (but does not accumulate) exergy in matter; (ii) the system gives off exergy to the surroundings in the form of heat and work, and receives exergy as chemical work $\dot{W}_{Ch}$ (by way of coupled chemical reactions) and useful radiation energy $\dot{Q}_u$ (by way of photosynthesis). Such contributions are identified by separate terms, assumed to be positive when received by the control volume.

For such conditions, the terms in (16) may be collected to give the following two equivalent expressions for the exergy expenditure:

$$
\begin{aligned}
\dot{\psi} = &- d[N(\bar{u} - T\hat{s})]/dt & \text{(from depots)} \\
&- \Sigma_{\text{out}}\, \dot{n}(\bar{h} - T\hat{s}) & \text{(from net intake)} \\
&+ \dot{Q}_u & \text{(from radiation)} \\
&+ \dot{W}_{Ch} & \text{(from chemical work)} & (20)
\end{aligned}
$$

or

$$
\dot{\psi} = \underset{\text{removed work}}{(-\dot{W})} + \underset{\text{destruction}}{\dot{I}} \,. \tag{21}
$$

These expressions are further reduced when the conditions for use of (17) or (18) are fulfilled. Also, for a living system as a whole, $\dot{W}_{Ch} = 0$. Finally, when the external work is zero, (21) shows that the exergy expenditure is equal to the irreversibility power, as expected.

Other conditions, e.g. growth, can be treated similarly using (4.12) and (4.13).

8.3.2 Exergy efficiencies

To evaluate the quality of one process, a living system as a whole or parts thereof, we can introduce efficiencies for exergy-supplying processes analogous to (4.14) and (4.15):

$$\varepsilon = \text{exergy supply/exergy expenditure} \qquad (22)$$

and for exergy-expending processes, analogous to (4.16):

$$\varepsilon = \text{theoretical exergy expenditure/actual exergy expenditure.} \qquad (23)$$

Such measures of quality, having values in the range $0 \leqslant \varepsilon \leqslant 1$, are based on sensible and relevant definitions that often depend on the problem at hand and are therefore best illustrated by examples.

8.3.3 The irreversibility

The irreversibility power for a control volume process can be determined from the exergy balance ((14), (16) or (17)) when all other terms in the equation are known. For a steady process, this requires known thermodynamic states of inflow and outflow of matter, as well as known supplies of heat and work power. For an unsteady process, the temporal change of state of the control volume must also be known. The examples that follow illustrate this use of the exergy balance. It should be noted that the entropy balance ((5.6) or (5.24)) in these cases would normally give $\dot{\sigma}$ and thereby $\dot{I} = T\dot{\sigma}$ or $\dot{I}_0 = T_0\dot{\sigma}$, cp. (5.28), which in fact is the exergy balance.

There are also cases where detailed knowledge of the irreversible processes themselves permit direct calculation of the entropy production and the irreversibility. This is true for the processes described in Chapter 7, dealing with local non-equilibria (chemical reactions) and global, spatial (discrete or continuous) non-equilibria (transport processes). According to the theory of irreversible thermodynamics ((7.50) and (7.57)), we may then use

$$\dot{\sigma} = \Sigma \, X_j J_j; \quad \dot{I} = T\dot{\sigma}, \qquad (24)$$

where X_j denotes generalized forces and J_j the associated generalized

234 Expenditure of exergy

fluxes. Some of the following examples will illustrate this approach to the calculation of $\dot{I}$ or $\dot{I}_0$, after which the exergy balance can be used to calculate other unknowns in this equation.

8.3.4 Thermal and mechanical exergy transformation

In this section, we illustrate the use of the exergy balance of some preceding examples which deal with thermal and mechanical processes in media treated as a pure substance. Since the exergy balance is a result of a combination of the first and second laws, its use requires detailed knowledge of the processes.

Example 8.1

·For the classical example of a work-producing cyclic process (e.g. a thermal power station) shown in Figure 8.2, where all quantities in (1) and (2) are positive, (14) for the shown control mass gives

$$0 = \dot{Q}_H(1 - T_0/T_H) - \dot{Q}_L (1 - T_0/T_L) + \dot{W} - \dot{I}_0, \qquad (a)$$

where $\dot{I}_0 = T_0\dot{\sigma}$, and T_0 is the temperature of the surroundings. Clearly, (a) can be derived directly as (1) minus (2) multiplied by T_0. The exergy expenditure is

$$\dot{\psi} = \dot{Q}_H(1 - T_0/T_H), \qquad (b)$$

the exergy yield is $\dot{W}$, and the exergy efficiency ((22)) becomes

$$\varepsilon = \dot{W}/[\dot{Q}_H(1 - T_0/T_H)], \qquad (c)$$

which is seen to deviate from the energy efficiency $\eta = \dot{W}/\dot{Q}_H$. Knowledge of temperatures, heat and work powers can be used to determine $\dot{I}_0$ from (a).

Example 8.2

For the subject analysed in Example 4.2, delivering a work power of 60 W while giving off a heat power of 300 W, the exergy expenditure is given by the two first terms in (20), or by (21),

$$\dot{\psi} = - \dot{W} + \dot{I}, \qquad (a)$$

provided the body temperature is chosen as the reference.

For lack of further information, we use the approximation

$((5.34))$, $\dot{I} \approx -\dot{Q}$, which is equivalent to assuming the left-hand side of (5.27) to be negligible. Then (a) gives

$$\dot{\psi} \approx 60 + 300 \approx 360 \text{ W}, \tag{b}$$

which is equal to the total energy expenditure. The exergy efficiency is therefore the same as η-total in Example 4.2.

If, instead, we use the general form $((14))$, with the temperature of the surroundings as reference (e.g. $T_0 = 10\,^\circ\text{C} = 283$ K, in place of $T = 37\,^\circ\text{C} = 310$ K), we obtain

$$\dot{\psi} = -\dot{Q}_H(1 - T_0/T) - \dot{W} + \dot{I}, \tag{c}$$

which, using the above approximation for the irreversibility, gives

$$\dot{\psi} \approx 300 \times (1 - 283/310) + 60 + 300 \approx 386.1 \text{ W}. \tag{d}$$

The increase in exergy expenditure, of the form (6), is equal to the work that could have been delivered by a Carnot machine working between heat reservoirs at T and T_0 respectively. While this result is relevant from the point of view of the universe as a whole, it is irrelevant from the point of view of bioenergetics, since living systems have no mechanisms for converting heat into work.

Example 8.3

For the isobaric heating of the steady air flow in Examples 4.8 and 5.6 (Figure 4.6, without the addition of water vapour), the exergy balance $((16))$, based on the tissue temperature T_t, reduces to

$$\dot{m}\, c_p\, [(T_2 - T_1) - T_t \ln (T_2/T_1)] = -\dot{I}, \tag{a}$$

which corresponds to (b) in Example 5.6, giving the numerical result

$$\dot{I} = 5.08 - 4.8 = 0.28 \text{ W}. \tag{b}$$

The left-hand side of (a), which represents the increase in exergy of the control volume (air), is seen to be exactly the negative of the irreversibility. This amounts to 6% of the supplied heat power and is due to the transfer of heat over a finite temperature difference. In this case, the irreversibility associated with heat transfer is relevant for the biological system.

Note that a direct calculation of the entropy production $\dot{\sigma}$ from expressions such as (5.20) or (7.10), and thereby $\dot{I} = T_t\dot{\sigma}$, is, strictly, not possible here. This is so because it is not known which

part of the heat power is transferred at which temperature difference. As an approximation, however, we can use the mean
temperature of the air $T_m = (T_1 + T_2)/2 = 291.5$ K, which, with
$T_t = 310$ K and $\dot{Q} = 4.8$ W from (c) in Example 4.8 substituted into
(5.20), gives

$$\dot{I} = T_t \dot{\sigma} = \dot{Q}\,(1 - T_m/T_t) = 4.8 \times (1 - 291.5/310) = 0.287 \text{ W.} \quad \text{(c)}$$

This estimate is quite accurate since the change in air temperature is
relatively small.

8.4   Chemical–thermal–mechanical processes

In the following sections, we re-examine selected processes from
Chapters 4 and 5 on the basis of the exergy balance in order to determine
exergy expenditure, exergy efficiency and irreversibility. Where possible,
the latter quantity is also calculated directly from the theory of Chapter 7
for non-equilibrium processes.

8.4.1   Chemical reactions and photosynthesis

For chemical reactions in biological systems, the exergy balance
reduces to (18). The negative of the left- or the right-hand side of (18)
then represents the exergy expenditure under normal conditions. So when
the reaction Gibbs function and the reaction rates are known for each of
the chemical reactions the exergy expenditure can be determined. In the
case of photosynthesis, the right-hand side of (18) is modified by the
addition of the supplied useful radiation power, as shown in (17).

Example 8.4

In Examples 4.15 and 5.7, the energy expenditure and the irreversibility was calculated to be $\dot{E} = 71.7$ W and $\dot{I} = 74.0$ W at a glucose
consumption of 389 g/d. We now wish to calculate the exergy
expenditure.

For the condition at rest ($\dot{W} = 0$), the exergy expenditure is
calculated from (20), which, with (b) in Example 5.8, gives

$$\dot{\psi} = -\dot{n}_G\,\Delta \hat{g}_r = 0.025 \times 2960 = 74.0 \text{ W.} \quad \text{(a)}$$

In this case, the ratio of $\dot{\psi}$ to $\dot{E}$ is equal to the ratio of $\Delta \hat{g}_r$ to $\Delta \hat{h}_r$.

Example 8.5

For the green leaf discussed in Examples 4.19 and 5.11, (17) reduces
to

$$\xi\, \Delta\hat{g}_r = \dot{Q}_u - \dot{I}, \tag{a}$$

and (20) and (21) reduce to

$$\dot{\psi} = -\,\xi\, \Delta\hat{g}_r + \dot{Q}_u = \dot{I}, \tag{b}$$

where $\xi = dN_G/dt$, averaged over the eight-hour period, can be
calculated from (d) in Example 4.19.

Typically, however, the determination of the contributions in (a)
and (b) requires separate use of the first and second laws, i.e. (b) in
Example 4.19 and (b) in Example 5.11.

From the above, and with values from (c) and (d) in Example
5.11, we can conclude that, for the eight-hour period in question,
the bound energy is 288 J, while the bound exergy is only 235 J,
since the process has required an exergy expenditure of 53.0 J
because of irreversibilities.

For the universe as a whole, the irreversibility is much larger since
an unused radiation power $\dot{Q}_a = \dot{Q}_s - \dot{Q}_u = 49$ mW/cm^2, given off
by the sun ($T_H \approx 6000$ K), is received by the leaf and rejected to the
surroundings ($T_0 = 298$ K). This gives the contribution

$$\dot{Q}_a\,(1 - T_0/T_H), \tag{c}$$

which should be included on the right-hand side of (a). For an
eight-hour period, (c) gives the additional irreversibility

$$I' = 8 \times 3600 \times 0.049 \times 10\,(1-\,298/6000) = 13\,410 \text{ J}, \tag{d}$$

which should be compared to $I = 53$ J. It can be argued that the
contribution (d) is irrelevant from a bioenergetic point of view,
since this contribution – and many other very large contributions –
will exist regardless of the presence of the leaf.

8.4.2 Coupled chemical reactions

The case of two coupled reactions treated in Sections 4.5.5 and
5.4.2, and exemplified by the combustion of glucose and the synthesis of
ATP (Figure 4.14), will now be re-examined.

Elimination of $\dot{Q}$ between energy and entropy balances in (5.35)–(5.38) and introduction of irreversibilities $\dot{I} = T\dot{\sigma}$, as well as Gibbs functions according to (5.30), yield exergy balances of the form (18) with $\dot{Q}_u = 0$. Since $\dot{I} \geq 0$, the exergy balances can be rewritten as the following conditions, which must all be fulfilled:

$$\dot{I} = -\xi_1 \Delta \hat{g}_{r,1} - \xi_2 \Delta \hat{g}_{r,2} \geq 0 \text{ (CV)}, \tag{25}$$

$$\dot{I}_1 = -\xi_1 \Delta \hat{g}_{r,1} - \dot{W}_{\text{Ch}} \geq 0 \text{ (CV)}_1, \tag{26}$$

$$\dot{I}_2 = -\xi_2 \Delta \hat{g}_{r,2} + \dot{W}_{\text{Ch}} \geq 0 \text{ (CV)}_2. \tag{27}$$

The inequality (25) gives an upper limit for the production of moles of ATP per mole of glucose:

$$\xi_2/\xi_1 \leq -\Delta \hat{g}_{r,1}/\Delta \hat{g}_{r,2}, \tag{28}$$

which may be compared to the values of this ratio, about 38 moles ATP per mole of glucose, as observed in living systems (see Example 8.6).

Note that $\dot{I}$ in (25) for the total spontaneous process is identical to that obtained from the relation ((7.5)) for local non-equilibrium processes which involve chemical reactions only. According to (26), $\dot{I}_1$ is less than would be expected from (7.5) because of the coupling, and $\dot{I}_2$ becomes positive, even though $\Delta \hat{g}_r > 0$, so that the total process (25) becomes spontaneous.

Example 8.6

For the oxidation of glucose in living cells (cp. Figure 4.14), the normally accepted yield is 38 moles of ATP per mole of glucose. We wish to compare this value with the theoretical maximal value and to estimate the irreversibility and heat loss of the process. For the reactions in question, at 37 °C, we use values from Appendix C.2, Examples 5.8 and 5.10.

$$\Delta \hat{g}_{r,1} = -2960 \text{ kJ/mol glucose}, \quad \Delta \hat{g}_{r,2} = 46 \text{ kJ/mol ADP} \tag{a}$$

$$\Delta \hat{h}_{r,1} = -2867 \text{ kJ/mol glucose}, \quad \Delta \hat{h}_{r,2} = 20 \text{ kJ/mol ADP.} \tag{b}$$

According to (28), the yield can be maximally

$$\xi_2/\xi_1 = -(-2960)/46 = 64.3 \text{ mol ATP/mol glucose,} \tag{c}$$

so that the process has an efficiency of $38/64.3 \approx 0.60$ for the transfer of useful energy.

The total irreversibility is obtained by (25) as

$$\dot{I}/\xi_1 = -(-2960) - 38 \times 46 = 1212 \text{ kJ/mol glucose,} \tag{d}$$

while the energy supplied as heat is found by (5.35) as

$$\dot{Q}/\dot{\xi}_1 = -2867 + 38 \times 20 = -2107 \text{ kJ/mol glucose.} \tag{e}$$

It should be noted that while the difference in reaction enthalpy and Gibbs function is vanishing for the combustion of glucose, this is not the case for the ATP synthesis. Using the given values, reaction entropies can be calculated according to (5.30), but determination of $\dot{W}_{Ch}$, $\dot{I}_1$ and $\dot{I}_2$ in (26)–(27) requires knowledge of the partition of $\dot{Q}$ on $\dot{Q}_1$ and $\dot{Q}_2$. Assuming, e.g. $\dot{Q}_2/\dot{Q}_1 = 1$, as in (a) of Example 4.21, we obtain, using (e),

$$\dot{Q}_1/\dot{\xi}_1 = (1/2) \times (-2107) = -1053.5 \text{ kJ/mol glucose,} \tag{f}$$

$$\dot{Q}_2/\dot{\xi}_2 = -1053.5 \text{ kJ/mol glucose,} \tag{g}$$

and (4.70) gives

$$\dot{W}_{Ch}/\dot{\xi}_1 = \dot{Q}_1/\dot{\xi}_1 - \Delta\hat{h}_{r,1} = -1053.5 + 2867$$
$$= -1814 \text{ kJ/mol glucose.} \tag{h}$$

Then (26) gives

$$\dot{I}_1/\dot{\xi}_1 = 2960 - 1814 = 1146 \text{ kJ/mol glucose.} \tag{i}$$

This is almost identical to the removed heat of (f), which is to be expected since the reaction entropy in (5.37) is very small for glucose combustion. Hereafter we find the irreversibility for the ATP synthesis, from (27), as

$$\dot{I}_2/\dot{\xi}_1 = -38 \times 46 + 1814 = 66 \text{ kJ/mol glucose,} \tag{j}$$

and for this part of the total process, the exergy expenditure equals $\dot{W}_{Ch}$, and the exergy efficiency becomes

$$\varepsilon = (\dot{W}_{Ch} - \dot{I}_2)/\dot{W}_{Ch} = 1 - 66/1814 = 0.96, \tag{k}$$

which seems reasonable, although somewhat high.

8.4.3 Electrochemical processes. Maintenance of structures

The analysis of electrochemical processes in Sections 4.6 and 5.5, exemplified by active transport of sodium ions across a cell membrane (Figure 8.5), will now be re-examined. Elimination of heat powers

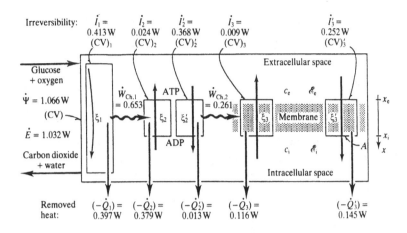

Fig 8.5 Example of active transport of Na^+ ions from intracellular space to extracellular space, in the direction of increasing electrochemical potential, and simultaneous opposing diffusive leakage in the direction of a decreasing electrochemical potential. This is Figure 4.15, now extended to include values for energy expenditure, transferred chemical work and removed heat (below), as well as exergy expenditure, exergy transfer and irreversibility (above).

between the energy balances ((4.76)–(4.80)) and entropy balances ((5.39)–(5.42)), yields exergy balances of the form of (18), from which the irreversibilities $\dot{I} = T\dot{\sigma}$ can be calculated:

$$-\xi_1 \, \Delta\hat{g}_{r,G} - \dot{W}_{Ch1} = \dot{I}_1 \; (CV)_1, \tag{29}$$

$$\xi_2 \, \Delta\hat{g}_{r,ATP} + \dot{W}_{Ch1} = \dot{I}_2 \; (CV)_2, \tag{30}$$

$$-\xi_2' \, \Delta\hat{g}_{r,ATP} - \dot{W}_{Ch2} = \dot{I}_2' \; (CV)_2', \tag{31}$$

$$-\xi_3 \, \hat{R}T \ln (c_e/c_i) - \xi_3 \, z(Na^+)\mathscr{F} \, (\mathscr{E}_e - \mathscr{E}_i) + \dot{W}_{Ch2} = \dot{I}_3, \; (CV)_3, \tag{32}$$

$$\xi_3' \, \hat{R}T \ln (c_e/c_i) + \xi_3' \, z(Na^+)\mathscr{F} \, (\mathscr{E}_e - \mathscr{E}_i) = \dot{I}_3', \; (CV)_3'. \tag{33}$$

For stationarity ($\xi_2' = \xi_2$ and $\xi_3' = \xi_3$), the sum of (29)–(30) gives the total irreversibility, which here is also the total exergy expenditure:

$$\dot{\psi} = -\xi_1 \, \Delta\hat{g}_{r,G} = \dot{I} \; (CV). \tag{34}$$

As expected, this quantity is solely related to the combustion reaction of glucose.

It is straightforward to define an exergy efficiency for each of the four processes in $(CV)_1$–$(CV)_3$ in Figure 8.5, using (29)–(32):

$$\varepsilon_1 = \dot{W}_{Ch1}/(\dot{I}_1 + \dot{W}_{Ch1}), \tag{35}$$

$$\varepsilon_2 = (\dot{W}_{Ch1} - \dot{I}_2)/\dot{W}_{Ch1}, \tag{36}$$

$$\varepsilon_2' = \dot{W}_{Ch2}/(\dot{I}_2' + \dot{W}_{Ch2}), \tag{37}$$

$$\varepsilon_3 = (\dot{W}_{Ch2} - \dot{I}_3)/\dot{W}_{Ch2}. \tag{38}$$

For $(CV)_3'$, there is no yield but only a tendency to break a spatial structure by diffusive leakage of Na^+ ions through the membrane. For this process, there is no efficiency ($\varepsilon = 0$). For the total process of (CV), the yield can be measured in terms of the capacity of the system to maintain the spatial structure. Here, the structure is characterized by the differences in concentration and electrical potential across the membrane. In general, *maintenance of structure* can be expressed quantitatively as the product of a generalized potential difference and a generalized flux. In the present case, this quantity equals the left-hand side of (33), or the irreversibility $\dot{I}_3'$ associated with the structure-breaking leakage process. On this basis, the exergy efficiency of the total process can be defined as

$$\varepsilon = \dot{I}_3'/\dot{\psi} = \dot{I}_3'/\dot{I}. \tag{39}$$

Example 8.7

We wish to calculate the total irreversibility and its partition among the processes described in Examples 4.21 and 5.12 for the transport of Na^+ ions, Figure 8.5. In addition, we seek the exergy efficiency for the total system.

Taking values from Examples 4.21, 5.8, 5.10 and Appendix C.2, (29)–(33) gives

$$\dot{I}_1 = 0.360 \times 10^{-3} \times 2960 - 0.653 = 1.066 - 0.653 = 0.413 \text{ W}, \text{ (a)}$$

$$\dot{I}_2 = -13.7 \times 10^{-3} \times 46 + 0.653 = 0.024 \text{ W}, \tag{b}$$

$$\dot{I}_2' = 0.629 - 0.261 = 0.368 \text{ W}, \tag{c}$$

$$\dot{I}_3 = -(0.001/60)\,[8.314 \times 310 \times (145/12) + 1 \times 96\,500 \times 0.090]$$
$$+ 0.261 = -0.252 + 0.261 = 0.009 \text{ W}, \tag{d}$$

$$\dot{I}_3' = 0.252 \text{ W}. \tag{e}$$

The first term in (a) is the sum of (a)–(e) so that (34) gives

$$\dot{\psi} = \dot{I} = 1.066 \text{ W}, \tag{f}$$

and from (39), with (e) and (f), we obtain the total exergy efficiency

$$\varepsilon = 0.252/1.066 = 0.24. \tag{g}$$

Efficiencies for the individual processes are obtained from (35)–(38),

$$\varepsilon_1 = 0.61; \quad \varepsilon_2 = 0.44; \quad \varepsilon'_2 = 0.41; \quad \varepsilon_3 = 0.96. \tag{h}$$

Figure 8.5 gives an overview of all energy and exergy powers.

Example 8.8

We wish to determine the irreversibility for $(CV)'_3$ in Figure 8.5, directly on the basis of the diffusion process of Na^+ ions from the extracellular to the intracellular space.

We assume that the diffusion flux $J(Na^+)$ in the x direction is constant over the area A (Figure 8.5) so that the total transport of sodium ions becomes

$$\xi_3 = J(Na^+)\, A. \tag{a}$$

In a differential layer of area A and thickness dx, the irreversibility is

$$d\dot{I}'_3 = T\,(\dot{\sigma}/\gamma)\, A\, dx, \tag{b}$$

which by integration over the thickness of the membrane gives the total irreversibility

$$\dot{I}'_3 = T \int_{X_e}^{X_i} (\dot{\sigma}/\gamma)\, A\, dx. \tag{c}$$

We substitute (7.21), the first term of which is zero for isothermal conditions, into (c) and using (a) also, we obtain

$$\dot{I}'_3 = -\xi_3 \int_{X_e}^{X_i} [d\bar{\mu}\, (Na^+)/dx]\, dx = \xi_3\, (\bar{\mu}_e - \bar{\mu}_i). \tag{d}$$

Assuming ideal mixtures and a vanishing pressure difference, the change in electrochemical potential in (d) can be calculated by integration of the last two contributions in (7.27):

$$\dot{I}'_3 = \xi_3\, [\hat{R}T \ln (c_e/c_i) + z\, (Na^+)\, \mathscr{F}\, (\mathscr{E}_e - \mathscr{E}_i)]. \tag{e}$$

The same result is obtained by using (3.136) with concentration as the activity measure, i.e. (3.125) with $y = 1$ for ideal conditions.

Result (e) is identical to (33), which, with the given assumptions of ideal conditions, was obtained from the first and the second laws. But this is not surprising since the same laws are the basis for the description of non-equilibrium processes in Chapter 7.

8.5 Model of a biological system

The analysis of the simplified model of a biological system in Figure 4.16, cp. Sections 4.7 and 5.6, can now be completed by the appropriate exergy balances. The irreversibility for each of the isothermal processes is determined by multiplication of the entropy balances, i.e. (5.45)–(5.49), by the absolute temperature and by subtracting the corresponding energy balance, i.e. (4.90), (4.91), (4.94), (4.98) and (4.99) respectively. This gives, on introduction of reaction Gibbs functions,

$$-\xi_1 \, \Delta \hat{g}_{r1} - (\dot{W}_{\mathrm{Ch2}} + \dot{W}_{\mathrm{Ch3}} + \dot{W}_{\mathrm{Ch3}} + \dot{W}_{\mathrm{Ch5}}) = \dot{I}_1 \, , \tag{40}$$

$$- (\xi_2 - \xi_2') \, \Delta \hat{g}_{r2} + \dot{W}_{\mathrm{Ch2}} = \dot{I}_2, \tag{41}$$

$$- (\xi_3 - \xi_3') \, \Delta \hat{g}_{r3} + \dot{W}_{\mathrm{Ch3}} = \dot{I}_3, \tag{42}$$

$$- (\xi_4 - \xi_4') \, \Delta \hat{g}_{r4} + \dot{W}_{\mathrm{Ch4}} = \dot{I}_4, \tag{43}$$

$$- (\xi_5 - \xi_5') \, \Delta \hat{g}_{r5} + \dot{W}_{\mathrm{Ch5}} + \dot{W} = \dot{I}_5, \tag{44}$$

and from the sum of these, for the whole system,

$$-\Sigma \, \xi_j \, \Delta \hat{g}_{rj} + \dot{W} = \dot{I} \, . \tag{45}$$

Here, $\dot{I}_j \geqslant 0$ and $\dot{I} \geqslant 0$, so that relations (40)–(45) impose restrictions on the ratio between reaction rates for exergy yielding and exergy consuming reactions. For the simple case of two coupled reactions, this was illustrated in Example 8.6. Rewriting (45) in accordance with (21), the total exergy expenditure becomes

$$\dot{\psi} = - \Sigma \, \xi_j \, \Delta \hat{g}_{rj} = (-\dot{W}) + \dot{I} \, . \tag{46}$$

Among the special cases, we consider stationarity for all processes, so that $\xi_j = \xi_j'$, $j = 2, \ldots, 5$, and only the contribution $j = 1$ will now enter the summation in (45) and (46). It follows that the total expenditure of

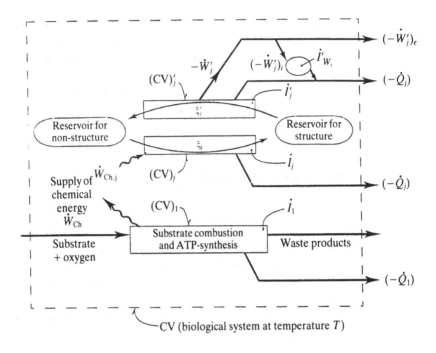

CV (biological system at temperature T)

Fig 8.6 Model of a biological system with supply of energy and exergy $(CV)_1$ to fuel a structure-building process $(CV)_j$, together with the associated spontaneous structure-breaking process $(CV)'_j$ that can supply internal and external mechanical work.

energy and exergy is determined solely by the combustion of substrate, just as for the ATP synthesis described in Section 8.4.2.

Further analysis of the biological model in Figure 4.16, including the partition of irreversibility among the individual processes, and quantification of structures, requires the separation of each of processes 2–5 into structure-building and structure-breaking processes, cp. Table 4.2. This is illustrated schematically in Figure 8.6, which shows a combustion process, with supply of energy and exergy, and one representative of the processes 2–5 with its associated reservoirs for structure and non-structure. It can be seen that Figure 8.6 corresponds to Figure 8.5, with the exception of the ATP reactions, which in Figure 8.6 proceed as part of the combustion process in $(CV)_1$.

We can use the special case of stationary processes in Figure 8.6 to summarize the important elements in a bioenergetic analysis. The first and the second laws give, for $(CV)_1$, $(CV)_j$ and $(CV)_j'$,

$$\xi_1 \Delta\hat{h}_{r1} = \dot{Q}_1 - \dot{W}_{Ch}; \quad \xi_1 \Delta\hat{s}_{r1} = \dot{Q}_1/T + \dot{\sigma}_1, \tag{47}$$

$$\xi_j \Delta\hat{h}_{rj} = \dot{Q}_j + \dot{W}_{Chj}; \quad \xi_j \Delta\hat{s}_{rj} = \dot{Q}_j/T + \dot{\sigma}_j, \tag{47}$$

$$-\xi_j' \Delta\hat{h}_{rj} = \dot{Q}_j' + \dot{W}_j'; \quad -\xi_j' \Delta\hat{s}_{rj} = \dot{Q}_j'/T + \dot{\sigma}_j', \tag{49}$$

where $\xi_j = \xi_j'$, and $\dot{W}_{Ch} = \Sigma \dot{W}_{Chj}$, $j = 2,3,\ldots$, denotes energy consuming processes, of which only some supply internal or external work, $(-\dot{W}_j')_i + (-\dot{W}_j')_e = (-\dot{W}_j')$. The exergy balance ((18)) gives

$$\xi_1 \Delta\hat{g}_{r1} = -\dot{W}_{Ch} - \dot{I}_1, \tag{50}$$

$$\xi_j \Delta\hat{g}_{rj} = \dot{W}_{Chj} - \dot{I}_j, \tag{51}$$

$$-\xi_j' \Delta\hat{g}_{rj} = \dot{W}_j' - \dot{I}_j'. \tag{52}$$

The energy expenditure of each process is the negative of the left- or the right-hand side of the first equation in (47)–(49), while the exergy expenditure is the negative of the left- or the right-hand sides of (50)–(52).

Exergy efficiencies of each process can be defined as in (35)–(38), while an exergy efficiency for the maintenance of structure for the total system, cp. (39), can be defined as

$$\varepsilon = [\Sigma(\dot{I}_j' - \dot{W}_j')]/(-\xi_1 \Delta\hat{g}_{r1}). \tag{53}$$

Here, the denominator is the total exergy expenditure $\dot{\psi}$, in accordance with (46). The numerator is the sum of all delivered internal and external work powers $(-\dot{W}_j')$ plus the sum of products of potential differences and fluxes for structure-breaking processes. The latter is also the sum of irreversibilities $\dot{I}_j'$ associated with these processes. All terms in the numerator represent energy that has been made useful. The magnitude of internal and external mechanical work powers can be estimated or observed. The power associated with structure-breaking processes can, in principle, be estimated as described in Section 8.4.3 and further illustrated in the following Example 8.9. This power represents the minimal exergy expenditure required to maintain the structure.

Example 8.9

A normal subject (weight 70 kg, volume $\approx 70\,l$, volume of active tissue $\approx 40\,l$, and surface area $\approx 1.7\,m^2$) has an energy expenditure at rest of $\dot{E} = 71.7 \approx 70$ W. This corresponds to the combustion of ≈ 389 g of glucose per day (cp. Examples 4.15 and 5.7), and the corresponding exergy expenditure has been calculated as $\dot{\psi} = 74.0$ W (Example 8.4). Here we show an approach to estimate the minimal exergy requirements for maintaining the structure of this system and we compare the estimate with the observed total exergy expenditures, the ratio being an overall exergy efficiency. We caution that the estimates are based on fragmentary knowledge.

In the following analysis, we include only important processes for the maintenance of structures, i.e. movement of blood and gut contents, ion pumps, and synthesis of protein molecules. The body is assumed to be isothermal ($T = 310$ K) and we can therefore neglect the irreversibilities associated with heat conduction over the small temperature differences that exist in the body.

Finally, we assume that none of the combustion processes takes place without coupling, i.e. without the formation of ATP, solely for the purpose of supplying the thermal energy that is required to maintain the body temperature above that of the surroundings. The necessary thermal energy is a natural byproduct of the inevitable irreversibilities associated with the structure-building and structure-breaking processes. It is possible that the control of body temperature is associated with changes in permeability of cell membranes and thereby with demand for pumping of ions and thus with related irreversibilities.

For each steady process, the minimal exergy expenditure can be determined by considering either the reversible structure-building or the irreversible structure-breaking process. The last approach reduces to the calculation of the irreversibility, which can be made either indirectly by way of the first and second laws, i.e. the exergy balance, or directly by way of entropy production according to the expressions derived in Chapter 7.

For the circulation of blood, for example, the reversible pump power is about 1.3 W (for 4.9 l/min at a pressure increase of 16 kPa, according to Example 4.9). For this adiabatic process, the reversible pump power equals the minimal structure-building exergy power and it also equals the irreversibility that is associated with the viscous dissipation observed as frictional pressure drop, the total contribution of which would be very difficult to estimate. Ventilation of lungs and propulsion of gut content can be estimated to be somewhat smaller and the power required to maintain a (minimal) tension in skeletal muscle at rest is presumably also small, so that maintenance of these mechanical structures can be estimated to require a minimal expenditure of exergy of about

Internal (reversible) mechanical work power ≈ 2–3 W.   (a)

For ion pumps, potential differences and fluxes can be estimated and their products give the structure-breaking irreversibilities. In order to estimate the total area of cell membranes in the body, we assume the total cell volume (about 40 l) to consist of spherical cells with diameters of 0.020 mm. This corresponds to a cell number of about 10^{13} and a total surface area of about 12 000 m^2. We now use the results for fluxes and potentials in (a), (b) and (n) of Example 7.5, cp. Figure 7.6, among which the most important contributions come from sodium and potassium ions. Use of (e) in Example 8.8, with $\xi'_j = A\,J_j$, gives for the passive flux of Na^+ into cells and the passive flux of K^+ out of cells,

$$\dot{I}' \ (Na^+ \ \text{ions}) = 12\,000 \times 249 \times 10^{-10} \times [8.314 \times 310 \times \ln\,(145/12)$$
$$+\ 96\,500 \times 0.09] = 4.52 \ \text{W}, \qquad\qquad (b)$$

$$\dot{I}' \ (K^+ \ \text{ions}) = 12\,000 \times 163 \times 10^{-10} \times [8.314 \times 310 \times \ln\,(139/3.7)$$
$$-\ 96\,500 \times 0.09] = 0.13 \ \text{W}. \qquad\qquad (c)$$

The electrical potential is the highest in the extracellular space so that the electrochemical potential difference is larger for Na^+ ions than for K^+ ions. The necessary supply of exergy to drive the ion pumps, and thereby to maintain the diffusive irreversibilities, is therefore largest for Na^+ ions. Among other contributions from ion pumps, the transport of Ca^{++} ions proceeds with a much smaller flux but a much larger electrochemical potential difference and this transport may contribute 1–2 W so that the total minimal expenditure of exergy to ion pumps may therefore be

Exergy to ion pumps ≈ 6 W.   (d)

For synthesis of proteins, RNA, etc., the exergy expenditure is determined mainly by the production of ATP that is consumed to form the peptide bonds in the amino acid chains that are continuously broken down. An estimate based on the structure-building processes requires knowledge of the number of ATP molecules required per peptide bond and the number of peptide bonds formed per unit of time in the body. Alternatively, the irreversibilities could be determined directly for the structure-breaking processes symbolized by the reaction ν_{FF} FF→F, i.e. process 3' in Figure 4.16. According to (7.3), multiplied by T, we have

$$\dot{I}'_3 = - A'_3\dot{\xi}_3' \; ; A'_3 = \hat{\mu}(F) - \nu_{FF} \; \hat{\mu}(FF), \tag{e}$$

where the reaction rate $\dot{\xi}'_3$ is the number of broken peptide bonds per unit of time for the whole body and A'_3 is the affinity, i.e. the degree of non-equilibrium, of the reaction, cp. (7.4). The affinity at equilibrium is $A^* = 0$ and we can derive the affinity at non-equilibrium by using (3.122) and (3.125), with $y = 1$ for ideal conditions:

$$A' = A - A^* = \hat{R}T \; \Sigma \; \nu_j \ln (c_j/c_j^*), \tag{f}$$

where c_j and c_j^* denote component concentrations at non-equilibrium and equilibrium, respectively, at T,p. In the case of (e), (f) gives

$$A'_3 = \hat{R}T \ln \{[c(F)/c^*(F)] \times [c^*(FF)/c(FF)]^{\nu_{FF}}\}. \tag{g}$$

Lack of experimental data precludes the use of (e) and (g) but other indirect estimates (Waterlow, 1984) suggest that the consumption of exergy for the maintenance of protein molecules in the body can be set at ≈ 10 W and a minimal value for the structure-breaking process at about half that value:

Synthesis of protein molecules ≈ 5 W. (h)

If we include other synthetic processes, e.g. glycogen from glucose (cp. process 2 in Figure 4.16), the sum of (a), (d) and (h) amounts to about

$$\Sigma \; \dot{I}' \approx 15 \text{ W}. \tag{i}$$

Hence, with a total exergy expenditure of 74 W this gives a net exergy efficiency of about

$$\varepsilon \approx 15/74 \approx 0.20. \tag{j}$$

This figure is considered to be a reasonable estimate. It may be compared to result (g) in Example 8.7, which, with the assumptions made there, gave $\varepsilon \approx 0.24$ for the sodium ion pump. However, each structure-building process may have different efficiencies for its subprocesses and it is not possible at this time to give more accurate accounts.

8.6   Concluding remarks about structures

As explained in the Introduction and reflected in Chapters 1–8, the purpose of this text has been to establish and test the foundations and tools for bioenergetic analyses. The theory is based on classical thermodynamics, supplemented with elements of irreversible thermodynamics.

Bioenergetic analyses are overall analyses and do not deal with details, for example with respect to intermediates in, and kinetics of, chemical reactions. The analyses take the existence of biological systems as given. The functions of the systems, and the processes within them, are characterized qualitatively and quantitatively through observations and measurements, both globally and locally, for example with the help of biochemical data.

The treatment is not concerned with the origin of biological systems, nor how they maintain their variety of structures in time and space. The energy and exergy analyses of processes and systems are based on the first and second laws, and mass conservation, and they merely confirm that the observed phenomena are in fact possible. In addition, we present estimates of the efficiency of biological systems to maintain the observed processes and structures and we find altogether reasonable values.

As discussed in the preceding section, cp. Figure 8.6, the observed processes can be divided into structure-building and structure-breaking processes, where 'structure' is defined in terms of generalized potential differences, e.g. spatial differences in electrochemical potential, in temperature or in pressure. During stationarity, the two types of processes balance to maintain the observed structures. Among examples of structures, we have mentioned the differences between electrical potential and concentration of Na^+ ions which are maintained over most cell membranes (Examples 4.21 and 8.7), where the membranes represent a considerable barrier to transport. Another example is the difference between intact and fractured macromolecules which requires the continuous synthesis of the macromolecules (cp. Figure 4.16), mainly proteins, which can be viewed as structures with finite life spans. The building process here involves the assembly of amino acids requiring the supply of energy and exergy, but there are normally no membranes, only concentration gradients to retard the process.

The structure-breaking processes (diffusion, friction, bond-breaking, certain chemical reactions, etc.) can be sufficiently well described by linear theory of irreversible thermodynamics for dissipative processes, cp. Chapter 7. These spontaneous processes can be summarily characterized by the dissipation of Gibbs energy (exergy destruction) that takes place.

On the other hand, the structure-building processes imply the existence of coupled processes, whereby a part of the Gibbs energy supplied (exergy) is used for structure building while the remaining part is dissipated. Such processes cannot be described by linear theories for irreversible processes, since linear processes cannot in general produce spatial structures over distances that exceed a few molecules. This fact is related to the phenomenon that a microscopic or macroscopic homogeneous system stays homogeneous, even though there is inflow and outflow of energy and a net supply of exergy, as long as the actual processes are linear, i.e. of small amplitude, say in electrochemical potential. The explanation is that random thermal fluctuations in such systems will rapidly dampen and smooth out the small non-uniformities and thus suppress any tendency to form spatial structures. This is not the case for large non-uniformities, or finite amplitudes, which are produced by non-linear processes that may therefore generate spatial structures.

The structure-building processes involve typically non-linear couplings between diffusive transport and several chemical reactions, often auto-catalytic reactions. Analysis of such processes leads to dynamic systems of non-linear differential equations, the solutions of which demonstrate that the processes can maintain non-uniform heterogeneous stationary states far from equilibrium. This fact was first recognized by Turing (1952). Since then, the dynamics and structure of non-linear systems in biology have been the subject of intensive research. This subject is not treated in the present text and we refer to the review by Olsen and Degn (1985), several papers in Degn, Holden and Olsen (1987) and in Markus, Mueller and Nicholis (1988), and, for an introduction, Field (1985).

Appendix A

Symbols

A	area, 24
A	Helmholtz function, 60
A	affinity, 193
A_j	variable in membrane transport, 210
a	activity, 66
a^0	standard activity, 66
B	extensive state property, 18
B_j	chemical symbol, 57
b	specific extensive state property, 18
b_j	mobility, 201
b^+, b^-	unidirectional mobility, 206
c	velocity of light, 34
c	heat capacity of incompressible matter, 52
c_j	concentration, 55
c_p	specific heat at constant pressure, 49
c_v	specific heat at constant volume, 49
CM	control mass, 16
CV	control volume, 17
D	diameter, 101
D	diffusion coefficient, 202
D_h	hydraulic diameter, 101
E	energy, 8
E	electric field strength, 25
E_a	activation energy, 194
$\dot{E}$	energy expenditure, 82
E_0	radiation energy, 42
$\mathscr{E}$	electrical potential, 25
e	specific energy, 18
F_{12}	view factor, 44
F_j	generalized intensive parameter, 9
$\mathscr{F}$	Faraday constant, 27
f	activity coefficient for mole fraction, 67
f	extent, $\varepsilon/\mathscr{V}$, 189

251

f	number of independent variables, 172
f_m	friction factor, 101
G	Gibbs function, 60
g	acceleration of gravity, 24
$\Delta \hat{g}_r$	reaction Gibbs function, 151
H	magnetic field strength, 29
H	enthalpy, 60
h	heat transfer coefficient, 40
h	specific enthalpy, 45
h_f^0	enthalpy of formation, 108
Δh_r	reaction enthalpy, 112
h_{fg}	enthalpy of evaporation, 95
I	electric current, 26
$\dot{I}$	irreversibility power, 141
δI	differential irreversibility, 141
i	electrical current density, 211
J	flux, 197
J^+, J^-	unidirectional flux, 202
K	force, 10
K	equilibrium constant, 185
KE	kinetic energy, 8
K_m	distribution coefficient, 210
k	heat conductivity, 40
k	number of components, 54
L	length, 40
L_{ik}	phenomenological coefficient, 213
l	length, 24
M	mass, 8
M	magnetization per unit volume, 29
$\hat{M}$	molar mass, 50
$\hat{M}_j$	molar mass, component j in mixture, 54
$\dot{M}_{ij}$	mass production in chemical reaction i, 56
m	number of permeating compounds, 178
$\dot{m}$	mass flow, 17
N	number of moles, 50
N_A	Avogadro's number, 27
$\dot{N}_{ij}$	mole production in chemical reaction i, 57
$\dot{n}$	mole flow, 57
$\dot{n}_e$	flow of elemental charges, 26
P	polarization per unit volume, 29
P_j	permeability coefficient, 199, 211
PE	potential energy, 8
p	pressure, 29
p	number of phases, 171
p^*	saturation pressure, 173
Δp_f	frictional pressure drop, 101
p_j	partial pressure, 55

Greek symbols

ν	frequency of light, 34
ν	kinematic viscosity, 101
ν_j	stoichiometric coefficient, 57
ξ	molar reaction rate, 58
ξ'	mass-based reaction rate, reverse molar reaction rate, 116
π	osmotic pressure, 68
ρ	density of mass, 17
ρ_j	mass concentration, 55
σ	force per unit area (stress), 29
σ	Stefan–Boltzmann constant, 42
$\dot{\sigma}$	entropy production, 141
σ_s	surface tension, 29
Φ	flux (through surface), 19
φ	potential energy, 18
χ	mole fraction, 54
ψ	exergy, 225

Special symbols (superscripts)

$(\bar{\ })$	partial specific state property, 62
$(\hat{\ })$	specific quantity per mole, 50
$(\ )^0$	standard state, 66
$(\tilde{\ })$	molar electrochemical state property, 70

Special symbols (subscripts)

$(\ )_j$	component j, 54
$(\ )_m$	mixture, 63
$(\ )_r$	reaction, radiation, 112
$(\ )_{rev}$	reversible, 9
$(\ )_u$	useful, 81

Appendix B

B.1 Dimensions and units

Quantity	Symbol	Dimension	Unit
Length*	l	L	m (metre)
Mass*	M	M	kg (kilogram mass)
Time*	t	T	s (second)
Force	K	MLT^{-2}	N (newton)
Energy	E	ML^2T^{-2}	J (joule)
Power	$Q, \dot{W}$	ML^2T^{-3}	J/s, W (watt)
Abs. temperature*	T	θ	K (degree Kelvin)
Rel. temperature	T	θ	°C (degree Celsius)
Entropy	S	$ML^2T^{-2}\theta^{-1}$	J/K
Electric current*	I	I	A (ampère)
El. charge	q	IT	C, As (coulomb)
El. potential	$\mathscr{E}$	$ML^2T^{-3}I^{-1}$	V (volt)
El. field strength	E	$MLT^{-3}I^{-1}$	V/m
El. polarization	P	$L^{-2}TI$	C/m^2
Dielectric displacement	D	$L^{-2}TI$	C/m^2
Dielectric constant	ε	$M^{-1}L^{-3}T^4I^2$	C/(V m)
Capacitance	C	$M^{-1}L^{-2}T^4I^2$	F, (C/V) (farad)
Inductance	L	$ML^2T^{-2}I^{-2}$	H, Vs/A (henry)
Magnetism	m	$ML^2T^{-2}I^{-1}$	Wb, Vs (weber)
Magnetic field strength	H	$L^{-1}I$	A/m
Magnetization	M	$L^{-1}I$	A/m
Magnetic induction	B	$MT^{-2}I^{-1}$	T, Wb/m^2 (tesla)
Magnetic flux	Φ	$ML^2T^{-2}I^{-1}$	Wb
Magn. permeability	μ	$MLT^{-2}I^{-2}$	Wb/m

*Fundamental SI units.

In addition to the fundamental units 1 m, 1 kg, 1 s, 1 K and 1 A, the fundamental unit of the amount of matter 1 mol is defined as: the amount of matter in a system which contains as many elemental parts (e.g. atoms, molecules or ions) as there are atoms in 0.012 kg C-12.

B.2 Conversion factors

Length	$1\,m = 10^6\,\mu m = 10^{10}\,\text{Å}$
	$1\,ft = 12\,inch = 0.3048\,m$
Mass	$1\,lbm = 0.4536\,kg = 453.6\,g$
	$1\,slug = 32.174\,lbm = 14.59\,kg$
Force	$1\,dyne = 10^{-5}\,N$
	$1\,kp = 9.807\,N$
	$1\,lbf = 4.448\,N$
Energy	$1\,erg = 10^{-7}\,J$
	$1\,kpm = 9.807\,J$
	$1\,kcal = 4187\,J\,(\text{IT calorie*})$
	$1\,Btu = 778.16\,ft \cdot lbf = 1055\,J$
Power	$1\,kcal/h = 1.163\,W$
	$1\,hp = 75\,kpm/s = 735.5\,W$
Pressure	$1\,bar = 10^5\,Pa = 100\,kPa$
	$1\,kp/cm^2 = 1\,at = 0.9807\,bar$
	$1\,atm = 760\,torr = 1.013\,25\,bar$
	$1\,lbf/in^2 = 1\,psi = 6895\,Pa$
Specific energy	$1\,kcal/kg = 4.187\,kJ/kg$
	$1\,Btu/lbm = 2.326\,kJ/kg$
Specific heat, entropy	$1\,Btu/lbm\text{-}R = 1\,kcal/kg\text{-}K = 4.187\,kJ/kg\text{-}K$
Temperature	$0\,°C = 273.15\,K = 491.69\,R = 32\,°F$
	$1\,°C = 1\,K = 1.8\,°F = 1.8\,R$
	$T_c\,°C = (T_f - 32) \times (5/9)$

*1 IT calorie $= 1\,cal_{IT} = 4.1868\,J$
$1\,cal_{15} \approx 4.1855\,J$, $1\,cal\,(\text{thermochem.}) = 4.184\,J$

B.3 Constants

Avogadro's constant	$N_A = 6.022 \times 10^{23}$ molecules/mole
Boltzmann's constant	$k = 1.3807 \times 10^{-23}\,J/K$
Dielectric constant (vacuum)	$\varepsilon_0 = 8.8542 \times 10^{-12}\,C^2 s^2/kg\text{-}m^3$
Mass of electron (at rest)	$m_e = 9.1094 \times 10^{-31}\,kg$
Elementary charge	$q_e = 1.6022 \times 10^{-19}\,C$
Faraday's constant	$\mathscr{F} = 96\,485\,C/mol$
Gas constant	$R = 8.314\,J/mol\text{-}K$
Permeability (vacuum)	$\mu_0 = 4\pi \times 10^{-7} = 1.2566 \times 10^{-6}\,kgm/C^2$
Planck's constant	$h = 6.626 \times 10^{-34}\,Js$
Stefan-Boltzmann constant	$\sigma = 5.670\,51 \times 10^{-8}\,W/m^2\text{-}K^4$
Standard acceleration of free fall	$g_n = 9.806\,65\,ms^{-2}$

Appendix C

Selected state properties

C.1

Enthalpy of formation $\hat{h}^0$ (kJ/mol), free energy of formation $\hat{g}^0$ (kJ/mol), absolute entropy $\hat{s}^a$ (J/mol-K) and heat capacity $\hat{c}_p^0$ (J/mol-K) for selected compounds at the standard state $(T_0, p_0) = (25\,°C,\ 1\,atm)$.

State of compound: g = gas, l = liquid, s = solid, aq = aqueous solution, ideal state at concentration 1 mol/l. For definitions, see Section 4.5.1.

Values have been taken from various tables (Schaefer and Lax, 1961; Lide, 1990) and have not been verified by the original publications.

Compound (state)	$\hat{h}^0$ (kJ/mol)	$\hat{g}^0$ (kJ/mol)	$\hat{s}^a$ (J/mol-K)	$\hat{c}_p^0$ (J/mol-K)
Carbon dioxide (g)	−393	−394	214	37
Carbon dioxide (aq)	−413	−386	121	
Acetic acid (l)	−487	−392	160	123
Acetic acid (aq)	−385	−396		
Acetate anion (aq)	−485	−369	87	
Ethanol (l)	−278	−175	161	114
Ethanol (aq)	−266	−182	222	
Carbonic acid (aq)	−699	−623	191	
Bicarbonate ion (aq)	−690			
Lactic acid (l)	−675	−518	192	
Lactic acid (aq)		−540		
Lactate anion (aq)	−687	−518		
Pyruvic acid (aq)		−489		
Pyruvate anion (aq)		−475		
α-D-glucose (s)	−1275	−910	211	219
α-D-glucose (aq)	−1264	−917	151	
Sucrose (s)	−2221	−1544	360	425
Sucrose (aq)		−1552		
Palmitic acid (s)	−882	−305	452	
Glycerol (l)	−669	−477		

Compound (state)	$\hat{h}^0$ (kJ/mol)	$\hat{g}^0$ (kJ/mol)	$\hat{s}^a$ (J/mol-K)	$\hat{c}_p^0$ (J/mol-K)
Glycerol (aq)		−488		
Triglycerol-palmitate (s)	−2457			
Hydrogen H_2 (g)	0	0	131	29
Oxygen O_2 (g)	0	0	205	29
Oxygen O_2 (aq)	−10	17	115	
Water (l)	−286	−237	70	75
Water (g)	−242	−229	189	34
Hydrogen ion (aq)	0	0	0	0
Hydroxyl ion (aq)	−230	−157	−11	−149
Oxonium ion H_3O^+ (aq)	−286	−237	70	75
Phosphoric acid (aq)		−1142		
Dihydrogen phosphate ion (aq)		−1130		
Hydrogen phosphate ion (aq)		−1089		
Phosphate ion (aq)	−1277	−1019	−222	
Carbamide (urea) (aq)	−319	−204	174	

C.2

Standard reaction enthalpy $\Delta \hat{h}_r^0$ and standard reaction Gibbs function $\Delta \hat{g}_r^0$ for selected reactions at the standard state $(T_0, p_0) = (25\,°C, 1\ atm)$. For further information see Blaxter (1989).

Reaction	$\Delta \hat{h}_r^0$ (kJ/mol)	$\Delta \hat{g}_r^0$ (kJ/mol)
ATP and glucose reactions:		
ATP → ADP + P (310 K, pH 7.0, 1 mmol Mg^{++}/l)	−20	−31
Glucose to ethanol, 298 K	−67	−235
Glucose to ethanol, 308 K	−66	
Glucose to lactic acid (298 K, pH 7.0)	−100	−198
Combustion reactions (kJ/mol substrate):		
Ethanol 293 K	−1371	
Acetic acid 293 K	−876	
Glucose 298 K, all components (aq)	−2870	
Glucose 310 K, all components (aq)	−2867	−2930
Lactose	−5646	
Palmitic acid, all components (aq)	−9982	−9791
Glycerol	−1659	
Triglycerides, average	−34 300	
Protein → urea, per gram, average	−17	
Protein → ammonia, per gram, average	−19	
Lactic acid	−1363	
Alanine	−1621	

Reaction	$\Delta \hat{h}_r^0$ (kJ/mol)	$\Delta \hat{g}_r^0$ (kJ/mol)
Alanine → urea	−1303	
Urea	−634	
Glutamic acid → urea	−1930	
Citric acid	−1986	
Proton-dissociation reactions:		
Neutralization by intracellular buffers	−25	
Acetic acid	−0.4	27
Glutamic acid		25
Water, 298 K	56	
Dissociation of strong electrolytes:		
Sodium chloride	−407	

References

Arpaci, V. S. (1986). Radiative entropy production, *Bull. Tech. Univ. Istanbul*, **39**, 291–305.

Arpaci, V. S. and Larsen, P. S. (1984). *Convection Heat Transfer*, Prentice-Hall Inc., Englewood Cliffs, New Jersey.

Atwater, W. O. and Benedict, F. G. (1903). *Metabolism of Matter and Energy in the Human Body*. Washington, DC: US Govt Printing Office (US Dep. Agric. Exp. Stn. Bull. **136**).

Benedict, F. C. and Milner, R. D. (1907). *Experiments on the Metabolism of Matter and Energy in the Human Body*. Washington, DC: US Govt Printing Office (US Dep. Agric. Exp. Stn. Bull. **175**).

Benedek, G. B. and Villars, F. M. H. (1974). *Physics – with Illustrative Examples from Medicine and Biology*, Addison-Wesley Publ. Co., Reading, Mass.

Blaxter, K. (1962). *The Energy Metabolism of Ruminants*, Hutchinson, London (third Impression 1969).

Blaxter, K. (1989). *Energy Metabolism in Animals and Man*. Cambridge University Press.

Brennen, C. and Winet, H. (1977). Fluid mechanics of propulsion by cilia and flagella, *Annu. Rev. Fluid Mech.*, **9**, 339–98.

Brønsted, J. N. (1955). *Principles and Problems in Energetics*, Wiley (Interscience), New York.

Callen, H. B. (1960). *Thermodynamics*, John Wiley and Sons, Inc., New York.

Carathéodory, C. (1909). *Mathem. Ann.*, **67**, 355 (for a discussion, see: Tisza, L. (1966). *Generalized Thermodynamics*, MIT Press, Cambridge, Mass.).

Degn, H., Holden, A. V. and Olsen, L. F. (eds.) (1987) *Chaos in Biological Systems*. NATO ASI, Series A: Life Sciences, vol. 138, Plenum Press, New York.

de Groot, S. R. and Mazur, P. (1962). *Non-equilibrium Thermodynamics*, North-Holland, Amsterdam.

Einstein, A. (1905). Über die von der molekularkinetische Theorie der Wärme geförderte Bewegung von in Ruhenden Flüssigkeiten suspendierten Teilchen, *Ann. Physik*, (4), **17**, 549.

Field, R. J. (1985). Chemical organization in time and space, *Amer. Scientist*, **73**, 142–50.

Fox, R. W. and McDonald, A. T. (1985). *Introduction to Fluid Mechanics*, (3rd edn), John Wiley and Sons, New York.

Garby, L. (1957). Studies on transfer of matter across membranes with special reference to the isolated human amniotic membrane and the exchange of amniotic fluid, *ACTA Physiologica Scandinavica*, **40**, Suppl. 137.

Garby, L. and Larsen, P. S. (1984). Principles and problems of human energy exchange, *Amer. J. Physiol.*, **247**, R255–72.

Gibbs, J. W. (1948). *Collected Works*, Yale Univ. Press.

Giedt, W. H. (1971). *Thermodynamics*, van Nostrand Reinhold Co., New York.

Gillies, R. J. (ed.) (1993). *Magnetic Resonance in Physiology and Medicine*, Academic Press, New York.

Glansdorff, P. and Prigogine, I. (1971). *Thermodynamic Theory of Structure, Stability and Fluctuations*, Wiley-Interscience, London.

Guggenheim, E. A. (1929). The conceptions of electrical potential difference between two phases and the individual activitis of ions, *J. Phys. Chem.*, **33**, 842–9.

Hamburger, K., Møhlenberg, F., Randløv, A. and Riisgård, H. U. (1983). Size, oxygen consumption and growth in the mussel *Mytilus edulis*. *Mar. Biol.*, **75**, 303–6.

Harned, H. S. and Owen, B. B. (1958). *The Physical Chemistry of Electrolytic Solutions*, Reinhold Book Corp., New York.

Hatsopoulos, G. N. and Keenan, J. H. (1962). A single axiom for classical thermodynamics, *J. Appl. Mech.*, **29**, 293–9.

Hatsopoulos, G. N. and Keenan, J. H. (1965). *Principles of General Thermodynamics*, John Wiley and Sons, Inc., New York.

Jacobsen, S., Johansen, O. and Garby, L. (1985). A 24-m^3 direct heat-sink calorimeter with on-line data acquisition, processing, and control, *Am. J. Physiol.*, **249**, E416–32.

Jaffrin, M. Y. and Shapiro, A. H. (1971). Peristaltic pumping, *Annu. Rev. Fluid Mech.*, **3**, 13–36.

Jørgensen, C. B., Famme, P., Kristensen, H. S., Larsen, P. S., Møhlenberg, F. and Riisgård, H. U. (1986). The bivalve pump, *Mar. Ecol. Prog. Ser.*, **34**, 69–77.

Jørgensen, C. B. (1990). *Bivalve Filter Feeding: Hydrodynamics, Bioenergetics, Physiology and Ecology*, Olsen and Olsen, Fredensborg, Denmark.

Kiørboe, Th., Møhlenberg, F. and Hamburger, K. (1985). Bioenergetics of the planktonic copepod *Acartia tonsa*: relation between feeding, egg production and respiration, and composition of specific dynamic action, *Mar. Ecol. Prog. Ser.*, **26**, 85–97.

Kresse, H. (1985). *Handbook of Electromedicine*, Siemens AG, John Wiley and Sons, Inc., New York.

Lide, D. R. (1990). *CRC Handbook of Chemistry and Physics*. Boca Raton, CRC Press, Inc.

Lienhard, J. H. (1981). *A Heat Transfer Textbook*, Prentice-Hall Inc., Englewood Cliffs, New Jersey.

Markus, M., Müller, S. C. and Nicholis, G. (eds.) (1988). *From Chemical to Biological Organization*, Springer Verlag.

Meyerhof, O. (1924). *Chemical Dynamics of Life Phenomena*, Lippincott, Philadelphia, PA.

Moran, M. J. (1982). *Availability Analysis, A Guide to Efficient Energy Use*, Prentice-Hall Inc., Englewood Cliffs, NJ.

Olsen, L. F. and Degn. H. (1985). Chaos in biological systems, *Quart. Rev. Biophys.*, **18**, 165–225.

Pedley, T. J. (1977). Pulmonary fluid dynamics, *Annu. Rev. Fluid Mech.*, **9**, 229–74.

Pledge, H. T. (1959). *Science Since 1950*, Harper Torchbooks, TB 506.

Prausnitz, J. M. (1969). *Molecular Thermodynamics of Fluid-phase Equilibria*, Prentice-Hall, Inc., Englewood Cliffs, New Jersey.

Prigogine, I. and Defay, R. (1954). *Chemical Thermodynamics*, Longmans, Green and Co., London.

Riisgård, H. U. and Poulsen, E. (1981). Growth of *Mytilus edulis* in net bags transferred to different localities in a eutrophicated Danish fjord, *Marine Pollution Bulletin*, **12**, 272–6.

Rock, P. A. and Gerholdt, G. A. (1974). *Chemistry*, W. B. Saunders Co., Philadelphia.

Schaefer, K. and Lax, E. (1961). *Landolt-Boernstein Zahlenwerk und Funktionen*. Volume II, Section 4. Springer, Berlin.

Sonntag, R. E. and van Wylen, G. J. (1966). *Statistical Thermodynamics*, John Wiley and Sons, Inc., New York.

Takabatake, S., Ayukawa, K. and Mori, A. (1988). Peristaltic pumping in circular cylindrical tubes: a numerical study of fluid transport and its efficiency. *J. Fluid Mech.*, **193**, 267–83.

Truesdell, C. (1971). *The Tragicomedy of Classical Thermodynamics*, Intl Center for Mech. Sci., Courses and Lectures No. 70, Udine 1971, Springer Verlag.

Turing, A. M. (1952). The chemical basis for morphogenesis, *Phil. Trans. Roy. Soc., London*, Ser. **B**, **237**, 37–72.

Vogel, S. (1981). *Life in Moving Fluids, The Physical Biology of Flow*, Princeton University Press, Princeton.

van Wylen, G. J. and Sonntag, R. E. (1965). *Fundamentals of Classical Thermodynamics*, John Wiley and Sons, Inc., New York.

Waterlow, J. C. (1984). Protein turnover with special reference to man, *Q. J. Exp. Physiol.*, **69**, 409–38.

Weissmann, G. and Claiborne, R. (eds.) (1975). *Cell Membranes. Biochemistry, Cell Biology and Pathology*, HP Publ. Co., Inc., New York.

Welch, G. R. (1985). Some problems in the usage of Gibbs free energy in biochemistry, *J. Theor. Biol.*, **114**, 433–46.

West, J. B. (1991). *Physiological Basis of Medical Practice*, (12th edn), Williams and Wilkins, Baltimore.

Yang, W-J. (1989). *Biothermal Fluid Sciences, Principles and Applications*, Hemisphere Publ. Corp., New York.

Index